The DAY *the* AMERICAN REVOLUTION BEGAN

19 April 1775

WILLIAM H. HALLAHAN

Perennial
An Imprint of HarperCollins*Publishers*

Maps on pages xvi–xvii courtesy of
the National Park Service.

A hardcover edition of this book was published in 2000 by William Morrow,
an imprint of HarperCollins Publishers.

HarperCollins books may be purchased for educational, business, or sales promotional use. For information please write: Special Markets Department, HarperCollins Publishers Inc., 10 East 53rd Street, New York, NY 10022.

First Perennial edition published 2001.

Designed by Kellan Peck

The Library of Congress has catalogued the hardcover edition as follows:

Hallahan, William H.
The day the American Revolution began : 19 April 1775/
William H. Hallahan—1st ed.
p. cm.

ISBN 0-380-97616-1

1. United States—History—Revolution, 1775-1783. 2. United States—History—Revolution, 1775-1783—Campaigns. 3. United States—History—Revolution, 1775-1783—Causes. I. Title.

E209.H36 1999 98-47260
973'3.dc—21 CIP

ISBN 0-380-79605-8 (pbk.)

05 JT/RRD 10 9 8 7 6 5

THE DAY THE AMERICAN REVOLUTION BEGAN

19 April 1775

For Marion

By the rude bridge
that arched the flood,
Here once the embattled farmers stood,
Their flag to April's breeze unfurled,
And fired . . .
. . . the shot heard 'round the world.

—RALPH WALDO EMERSON

Contents

Illustrations

Grateful acknowledgment is made to the Franklin Collection, Yale University Library for permission to use the engraving of Joseph Galloway and to the Independence National Historic Park for permission to use all other portraits.

Note to the Reader

The day the American Revolution began was not the same day for everyone.

Historically, it began after dawn on Wednesday, April 19, 1775, in Lexington and Concord, Massachusetts. But for the residents of New York City the Revolution began on the following Sunday afternoon, April 23 at four P.M., when express rider Israel Bissel paced his tired horse into that explosive city, crying the news. Philadelphians learned of it the next day, Monday, April 24. Late at night on April 28, an express rider cantered into Williamsburg, Virginia, and roused sleeping residents, who were themselves on the verge of starting the American Revolution. Weeks later a ship would carry the news to the people of Georgia. And in London, it was not until May 28 that stunned cabinet members hastened out of the city by carriage to bring the news to King George III, who was weekending with his family in the royal palace at Kew.

This narrative, rather than dispassionately analyzing events and causes, follows the path of the news from city to city and country to country, into homes, taverns, farms, parliamentary offices, and palaces, and thereby rediscovers that intense emotional moment when word of Lexington and Concord abruptly tore open the lives of many diverse people.

Each life was changed in significantly different ways. While the news was heard with great dismay by John Harrower, an indentured servant in Fredericksburg, Virginia, who was struggling to bring his wife and children from Scotland, Benedict Arnold saw great personal opportunity in it and dared to reach for fame and glory. While deeply troubled Quakers such as Samuel Wetherill were struck suddenly by the realization that their determined pacifism would be regarded by both sides as betrayal, women such as Abigail Adams saw that any new freedom would be for men alone. And while many Loyalists such as Oliver DeLancey would try to fight back, only to be ruined—their great fortunes lost, their magnificent mansions torched, and they themselves permanently exiled—leaders at the other end of the political spectrum such as George Washington reacted by also risking all, hazarding everything, including their lives, to ride to Philadelphia and join the Revolution.

By peering into these individual lives one finds that many of the thrice-told tales are not true. The picture of George III as an insane tyrant is hardly accurate. The view of all Englishmen and the Parliament standing behind their king, unified against the damned rebels, does not square with the violent disputes that racked Lords and Commons, drawing rooms and taverns, throughout that country. Paul Revere never shouted "The Redcoats are coming," nor was he, as Longfellow reported, the express rider who roused Concord that night. There is even a legitimate question as to which shot should be deemed the shot heard 'round the world—the first one fired on the green at Lexington or, Emerson's choice, the later one at the North Bridge in Concord.

Nor were our founding fathers the giants of the earth that boosters such as Parson Weems painted. Mixed among men of stature were smugglers, slave owners, street thugs, arsonists, hack politicians, hypocrites, opportunists, self-servers, liars, spies, traitors, double dealers, and cowards. But rather than sullying their accomplishments, these flaws make their deeds even more impressive. For to defy a powerful king, to fight and freeze through an exhausting war, and to write the Declaration of Independence and the Constitution, they had to reach far beyond their normal capacities to become true heroes.

History is written not by the historians but by the people who lived it. And here they are, from colonial farmer to king—as varied a group as ever shared the same era, ordinary people made extraordinary by the great tidal wave from Lexington and Concord that swept over them. They were about to live through—and cause—some of the most dramatic, absorbing, even improbable, events in all of American and English history.

As you will see, the day the American Revolution began has a particularly modern resonance.

—W.H.H.

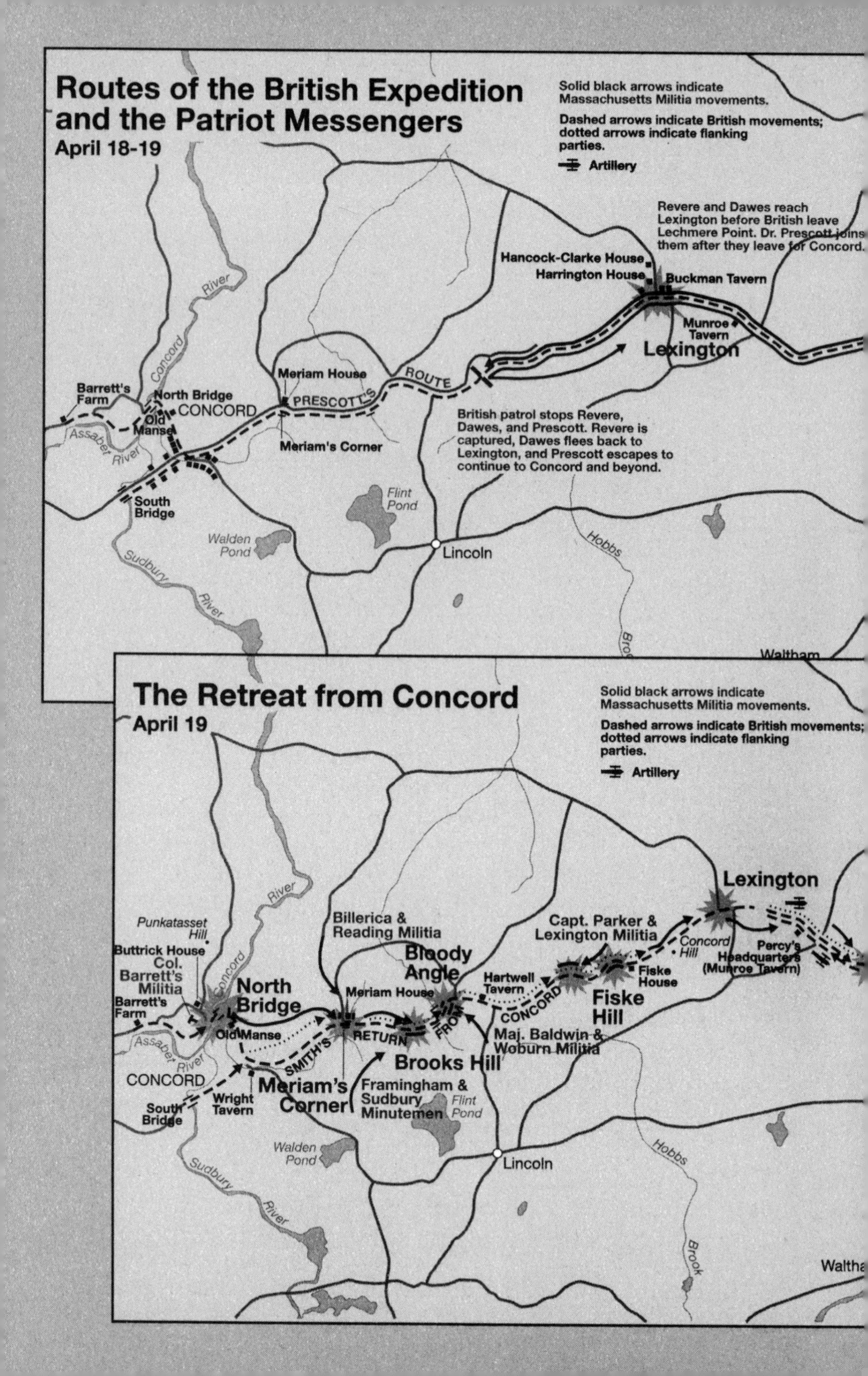

Routes of the British Expedition and the Patriot Messengers
April 18-19
Solid black arrows indicate Massachusetts Militia movements.
Dashed arrows indicate British movements; dotted arrows indicate flanking parties.
Artillery
Revere and Dawes reach Lexington before British leave Lechmere Point. Dr. Prescott joins them after they leave for Concord.
Hancock-Clarke House
Harrington House
Buckman Tavern
Munroe Tavern
Lexington
Concord River
Barrett's Farm
North Bridge
Old Manse
CONCORD
Meriam House
PRESCOTT'S ROUTE
Meriam's Corner
Assabet River
South Bridge
British patrol stops Revere, Dawes, and Prescott. Revere is captured, Dawes flees back to Lexington, and Prescott escapes to continue to Concord and beyond.
Flint Pond
Walden Pond
Lincoln
Sudbury River
Hobbs
The Retreat from Concord
April 19
Solid black arrows indicate Massachusetts Militia movements.
Dashed arrows indicate British movements; dotted arrows indicate flanking parties.
Artillery
Lexington
Punkatasset Hill
Billerica & Reading Militia
Capt. Parker & Lexington Militia
Concord Hill
Percy's Headquarters (Munroe Tavern)
Buttrick House
Col. Barrett's Militia
Barrett's Farm
North Bridge
Bloody Angle
Hartwell Tavern
Fiske House
Fiske Hill
Meriam House
Old Manse
SMITH'S RETURN FROM CONCORD
Brooks Hill
Maj. Baldwin & Woburn Militia
Assabet River
CONCORD
Meriam's Corner
Framingham & Sudbury Minutemen
Flint Pond
South Bridge
Wright Tavern
Walden Pond
Lincoln
Sudbury River
Hobbs Brook
Waltha

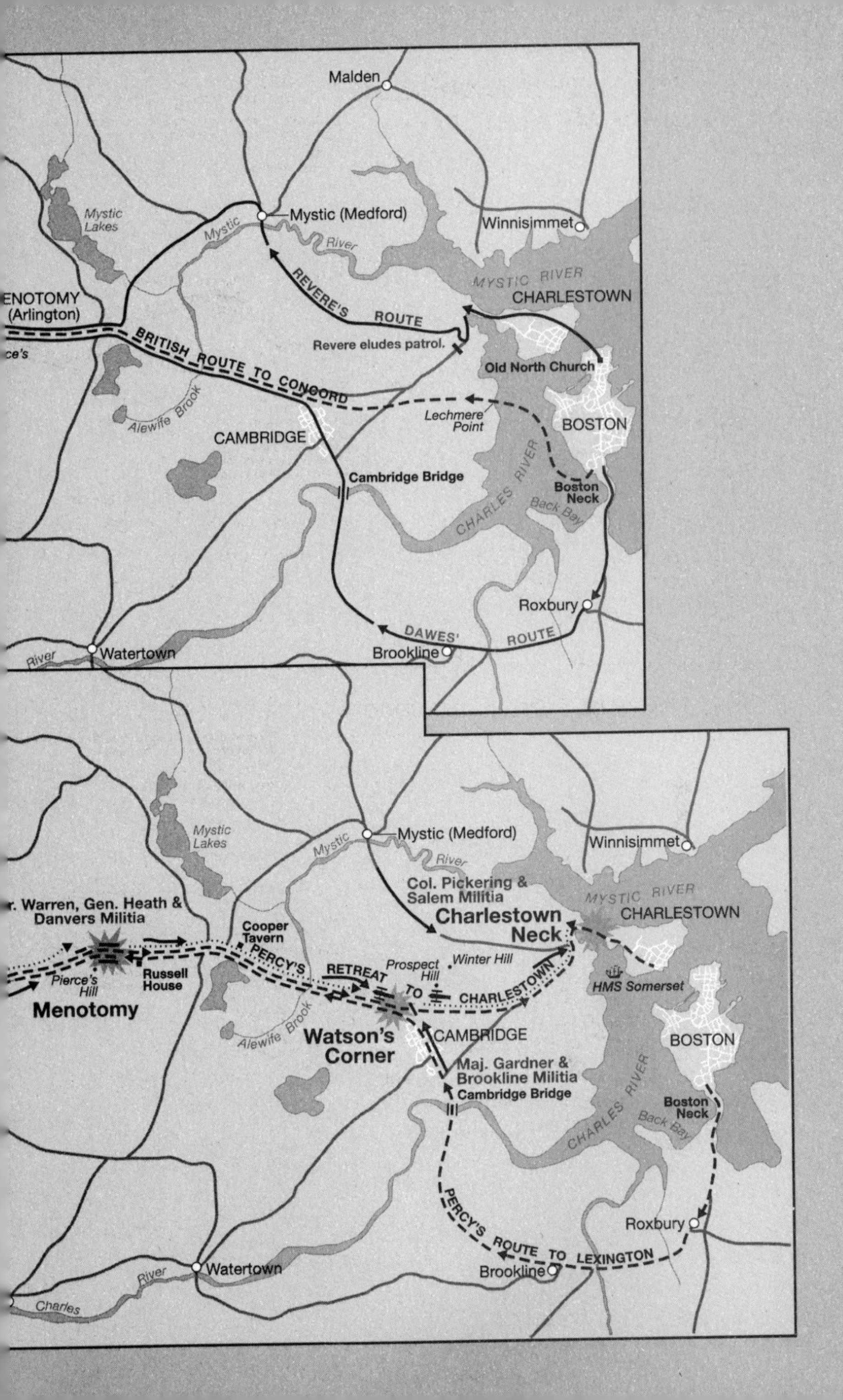

Malden
Mystic Lakes
Mystic (Medford)
Winnisimmet
Mystic
River
MYSTIC RIVER
CHARLESTOWN
ENOTOMY
(Arlington)
REVERE'S ROUTE
BRITISH ROUTE TO CONCORD
Revere eludes patrol.
Old North Church
Alewife Brook
Lechmere Point
BOSTON
CAMBRIDGE
Cambridge Bridge
CHARLES RIVER
Boston Neck
Back Bay
Roxbury
DAWES' ROUTE
Watertown
Brookline
River
Mystic Lakes
Mystic (Medford)
Winnisimmet
Mystic
River
Col. Pickering & Salem Militia
MYSTIC RIVER
CHARLESTOWN
Charlestown Neck
r. Warren, Gen. Heath & Danvers Militia
Cooper Tavern
PERCY'S RETREAT TO CHARLESTOWN
Prospect Hill
Winter Hill
Russell House
Pierce's Hill
HMS Somerset
Menotomy
Alewife Brook
Watson's Corner
CAMBRIDGE
BOSTON
Maj. Gardner & Brookline Militia
Cambridge Bridge
CHARLES RIVER
Boston Neck
Back Bay
PERCY'S ROUTE TO LEXINGTON
Roxbury
Watertown
Brookline
River
Charles

THE DAY THE AMERICAN REVOLUTION BEGAN

19 April 1775

Boston

On the morning of Tuesday, April 18, 1775, under bright skies and in cool spring temperatures, twenty British officers and sergeants, all mounted, all under the command of the mercurial Major Edward Mitchell of the 5th Foot, rode out of the British army stables in Boston. They traveled down the long thread of land called the Boston Neck, the only land access from the city to the mainland, then, turning north and west, they spread out along the road to Concord.

The preceding winter had been the mildest in New England history. The roads were dry underfoot, and their horses could cover ground quickly. The greenery had begun to grow a month early, and the trees were already budding out.

From their horses, these British soldiers looked at familiar scenes in a peaceful countryside. Farmers were out in their fields, plowing with oxen from first light to the last peep of sun. On the roads, drovers with switches and flailing hats were driving animals—everything from cattle to gabbling flocks of turkeys—to Boston's market. Everywhere, people were doing chores and errands in the urgent business of getting crops started. Countless eyes saw Major Mitchell and his mounted military detachment pass by.

Dr. Joseph Warren

Boston itself was even quieter on that April Tuesday. The port had been closed since the previous June 1 by order of His Majesty King George III until such time as the citizens reimbursed the East India Company for tea that had been dumped into Boston Harbor.

A closed port meant a dead city. Almost all commerce had ceased. No ships came or went. Most businesses were shut down: the rope walks, the sail lofts, the ships chandlers, all closed; the warehouses empty. Large groups of sullen, unemployed men stood about on the docks or moped in the taverns. The other colonies were sending food to ward off starvation. So far, the city had refused to pay for the tea.

Dr. Joseph Warren was in his Boston medical office but not attending to medical duties. He was spending the day sifting through bits and pieces of information that grew more ominous as the hours passed. For Dr. Warren was spying on General Thomas Gage and his troop movements.

A leading physician, a graduate of Harvard University, and the son of a leading Boston family, Dr. Warren was also a passionate Whig, a leading radical, and one of Samuel Adams's most valued lieutenants. He was also a leading member of the Massachusetts Committee of Safety, an extralegal executive body of the radicals that had charge of Massachusetts's huge militia.

He had achieved lasting fame among the radicals when, while Sam Adams was attending the First Continental Congress, he had written the Suffolk Resolves under Adams's instructions. Paul Revere had carried them to Philadelphia, where they, by providing the defiant tone which Adams pushed through, turned out to be the turning point of the First Congress.

Dr. Warren was still weighing an important piece of intelligence. There had been a change of assignment in the British garrison that indicated a potential military maneuver: On April 15, three days earlier, Gage had detached his elite troops—grenadiers and light infantry—from their regular duties in order to give them separate training. Lightly equipped, specially trained, and fast moving, these companies could be used to strike into the countryside and return with great speed. All told these units contained 700 men—just the right size for a hit-and-run strike. But when and where were they to be used?

Dr. Warren was one of only a handful of radicals still left in Boston. None were to be found in their shops or countinghouses, or haranguing each other in taverns such as the Green Dragon or the Bunch of Grapes, or in the popular British Coffee House. They had all heard rumors from London that the king's cabinet, out of patience, was demanding results from General Gage and that the king had ordered him to take sterner action—to confiscate illegal arms, to break up illegal assemblies, and to arrest the radical leaders, especially John Hancock and Sam Adams.

So on April 8, the radical politicians had prudently moved in large numbers out into the countryside, beyond the reach of Gage's long, red-coated arms. In fact, a number of radicals had defiantly attended an illegal Provincial Council meeting in Concord to select delegates to the equally illegal Second Continental Congress in Philadelphia. They also decided to remain safely outside Boston to watch from afar and await Gage's move.

Dr. Warren had even sent his own family away for their protection, and every military uniform that passed his office windows reminded him that by remaining in the city he was continually in danger of sudden arrest and confinement. But he couldn't leave yet. He still needed one piece of vital information.

Silversmith Paul Revere had also elected to stay in Boston. From his shop on Fish Street, near Hancock's Wharf, he was watching and waiting along with Dr. Warren, secretly gleaning hints and clues from informants all over the city. He, too, believed that the British garrison was on the verge of launching an operation.

To the angry British officials who knew him well, Paul Revere was primarily a post rider for the radical groups and a thorn in their sides. But he was much more than that.

He was a deeply committed radical ready for any assignment. A man with many skills—a gifted silversmith, a goldsmith, a pewtersmith, a maker of false teeth, and an engraver—he was also a political cartoonist, a writer, a newspaper propagandist, a rioter, and an occasional arm bender. The physically powerful Revere was more than Sam Adams's leading post rider. He was his most trusted lieutenant.

A gifted organizer as well, he had woven dozens of villages outside Boston into a network of post riders to carry alarms all over the colony. Most recently he had organized a band of thirty radicals—called the Committee of Observation—to spy on General Gage. The departure that morning of Major Mitchell and his twenty horsemen had not passed unnoticed.

By mid-afternoon, however, it seemed to Revere and Dr. Warren that this would only be another uneventful day. All was stubbornly quiet in the British barracks. But General Thomas Gage, commander in chief of all British forces in North America and also royal governor of Massachusetts, was far from inactive. He had conferred with Hugh, Lord Percy, colonel of the 5th Foot, and in the strictest secrecy revealed to him a plan of operations based on orders from England.

Later, in the Green Dragon Tavern, a young boy listened to the conversation of two British officers. What he heard sent him running to Revere.

In their shop on Queen Street, publishers Benjamin Edes and John Gill were struggling with a decision. As proprietors of the *Boston Gazette*, one of the most incendiary newspapers in all the colonies, they knew that they were in imminent danger of being flung into prison.

In fact, they needed do no more than look into the empty shop of their competitor and cohort Isaiah Thomas, the twenty-six-year-old publisher of the equally inflammatory *Massachusetts Spy,* to know how dangerous their position was.

Thomas had been given the most graphic possible warning to get out of town. His newspaper had made such savage attacks on the British army that every last man of the 47th Regiment, including a furious red-faced colonel and a full regimental band, had gathered in front of his print shop and warned him he was within inches of being tarred and feathered. It was hardly an idle threat. British soldiers had already tarred and feathered others.

As a result, two days earlier, on April 16, with aid sent by the omnipresent Dr. Warren and another dedicated radical, Timothy Bigelow, Isaiah Thomas had prudently packed up his family, his press and his type and carried them all out of Boston by the Charlestown ferry. His merrily philandering wife and their two children he left in Watertown, but his press was sent by wagon forty miles inland to Worcester and put in Bigelow's basement, far from the reach of British authorities in Boston.

It was surprising that the 47th had confined its threats to just Thomas's paper. Edes and Gill had been equally guilty. Under the tutelage and guidance of Sam Adams, Edes and Gill, those "foul-mouthed Trumpeters of Sedition," for years had slanted stories with unparalleled invective, fabricated events, distorted and suppressed the truth, and been as responsible as any other single agency for enflaming the colonists against everything British—Crown, Parliament, and army. More than even Thomas's *Spy,* the *Boston Gazette* was "crammed with treason," and was blamed by the Tories for rousing the people of Massachusetts—and those far beyond—to the pitch of hysterical rebelliousness. In fact, it was common tavern gossip that the members of the Boston tea party had blackened their faces and put on their Indian disguises in the offices of the *Gazette.* So outrageous was the treatment of the truth by Edes and Gill that James Rivington's Tory paper in New York, the *Gazetteer,* dubbed the *Gazette* "The Weekly Dung Barge."

The quandary for the two publishers was tormenting. In a city awash with rumors of imminent British army action, should they risk arrest by taking the time—several days at least—to pack their type boxes and printing press? Or should they abandon everything and depart immediately? They had to make a decision quickly. Tomorrow, they realized, might be too late.

* * *

In Charlestown, across the river from Boston, lived a mare by the name of Brown Beauty, who, among the hordes of horses around her, had the reputation for being uncommonly fast. By all accounts she was believed to be an excellent saddlehorse—big, strong, tireless—akin to a breed in England today called the Suffolk Punch. Within the next twelve hours she would gallop into the history books and bring lasting honor to her owner, John Larkin, deacon of the Charlestown Congregationalist Church.

In Roxbury, due south of Boston Neck, William Heath, a thirty-eight-year-old gentleman farmer with a passion for Whig politics and military history, strategy, and tactics, had warfare on his mind. Looking exactly like the country squire he was, a man of immense bulk, bald with kindly eyes and a mirthful manner, he had not one day's experience in combat. However, as he had confided to his close friend Dr. Joseph Warren, he was convinced that a fight between the colonists and the British army was not only inevitable but approaching, and his deep studies in military strategy had given him a plan that might make all the difference.

In Braintree, ten miles south of Boston, John Adams, already famous for his defense of the soldiers accused of the Boston Massacre, was putting his affairs in order prior to his trip to the Second Continental Congress in Philadelphia. The previous autumn his journey to the First Congress had been his first trip ever outside Massachusetts, and he had been deeply impressed with the men he had met from the other colonies. As a result, the idea of common cause and union, which had always seemed impossible to him, now seemed more plausible. The Second Congress would be the rostrum to explore the idea of union further.

In Concord, surrounded by freshly plowed farms twenty-two miles northwest of Boston, Reverend William Emerson was up and about, wondering if the British were going to strike at his community. For here the extralegal Provincial Council—the shadow government that had run the colony since the governor had shut down the established legislature more than a year earlier—had been meeting in Concord for several weeks before disbanding just the Friday before.

Meanwhile, Reverend Emerson's fellow townsman, the handsome

twenty-three-year-old Dr. Samuel Prescott, was planning to make the five-mile ride to Lexington that evening. He would be wearing an expensive suit of the highest fashion in order to make the best possible impression on a celebrated beauty by the name of Miss Lydia Mulliken. He had betrothal on his mind. With luck, he would not be on the road back to Concord until late that night. He had not the slightest inkling that by the following dawn he would become an American legend.

In Miss Mulliken's village, John Parker, farmer, mechanic, captain of the militia, born leader, and one of the most trusted men in the town, was not well. In fact, forty-six-year-old Parker was in the late stages of tuberculosis and had slept poorly the night before. Yet, with his extensive war experience, he was an ideal choice to head up the town's militia. Thin by nature and made thinner by disease, he faced a two-mile walk from his home to the village green, where the troops customarily mustered and trained. And two long miles home again.

On his farm outside the town of Dover, twelve miles southwest of Boston, militiaman and farmer Jabez Baker looked critically at his freshly plowed fields and a riot of birds. They dashed about in search of food for their newborn young. He decided he needed a proper scarecrow that the birds would really fear. What could that be? he wondered. What could he use? Setting his eye to watch for the right item, he could not have guessed what unusual object he would eventually find or what a durable legend it would create.

In the town of Lynnfield, twenty miles north of Boston, another young doctor, Martin Herrick, a graduate of Harvard, a member of his town's militia, and one of Paul Revere's volunteer express riders, would finish his medical rounds and house calls at day's end by riding fifteen miles back to his home in Medford.

In Watertown, ten miles from Boston, a post rider named Israel Bissel would, in a few hours, start the most remarkable ride of his life.

Back in Boston, by mid-afternoon Dr. Warren had accumulated more clues but still nothing definitive. He was fairly sure General Gage was prepared to make a strike against the colonists, he even felt he knew where, but he didn't know by what route or when.

Also, he was puzzled by the strange behavior of the British navy. Tethered under the sterns of the two warships in the harbor, HMS *Somerset,* sixty-four guns, and HMS *Boyne,* seventy guns, were the ships' longboats, customarily stowed on the decks of their ships. Floating there in the April sunlight, they could quickly be loosed for transporting troops by water.

Dr. Warren had to admit this might be a ruse to put off the watching radicals. After all, it was so obvious. But if it had been the work of Admiral Graves, the port's ranking naval officer and one of the most corrupt men in the British navy, then perhaps the longboats weren't decoys. The radicals believed Graves was too "unbelievably stupid" to be devious. That meant that the admiral had been ordered by the general to prepare for water transport of troops and so had slapped the boats into the water in full view of Boston, of Charlestown, even of Cambridge. But when? Those boats had been floating around the ships' sterns for days.

There were other indications but no hard facts. Even so, both Dr. Warren and Paul Revere were slowly becoming more sure that something was afoot. Everything pointed toward that very night.

By late afternoon, Revere decided to make a key move. He left his shop and went to see Robert Newman, an unemployed leather worker and dedicated patriot, who was the sexton of the Old North Church. After conferring with Newman, Revere then searched out Captain John Pulling, a vestryman of the same church, and after him, a friend and neighbor named Thomas Bernard. He told all three men to be on call for action that night. Then he met with two other Whigs, Joshua Bentley, who built boats for a living, and Thomas Richardson, and told them to be prepared as well.

In Boston, Dr. Benjamin Church was making his plans to leave the city.

In his early forties, a deacon's son, and a graduate of the London Medical College, Dr. Church was one of Boston's leading physicians. He was also known for his charitable works, such as giving free inoculations to the poor during smallpox epidemics. It was he who had been given the assignment to examine the battered body of Crispus Attucks, killed during the Boston Massacre.

But Dr. Church was more than a physician. A charter member of the Long Room Club, located over the Edes and Gill print shop, he was one of Samuel Adams's best propagandists and satirists, and as such had written the famous, long-remembered, and oft-repeated verse attack on Governor Francis Bernard:

Fop, witling, favorite stampman, tyrant, tool.
Or all those mighty names in one, thou fool!

Dr. Church's devotion to the radical cause was unflagging. He served with Dr. Warren on every crucial committee, including the powerful Committee of Safety, and was a member of the Provincial Congress. Dr. Church was one of the highest-ranking, most admired, and respected patriots in Massachusetts. Following Lexington, he would be appointed surgeon general of Washington's army. He was also one of Sam Adams's favorite companions.

Yet he was a disturbing puzzle to his fellow patriots such as Paul Revere, who looked with doubt on Dr. Church because of his notorious reputation as a womanizer who kept his mistress in high style; because, though not wealthy, he had built an expensive house in Raynham; because Revere had often watched Church dine with British officials; and because Church had been seen leaving General Gage's house making an effusive parting "like old friends."

Church's plans for leaving were twofold. Late that evening, he would hold a clandestine meeting with a major in the British army. Then he would bid farewell to his mistress.

In Lexington, at six o'clock in the evening, Elijah Sanderson, a cabinetmaker and militiaman, was surprised to see a group of mounted British officers and sergeants loitering on the roadway near his town with darkness coming. He was suspicious enough to carry his news to the Buckman's Tavern in Lexington. But first he stopped to get his musket and ammunition box.

Solomon Brown, eighteen, and also a Lexington militiaman, saw the same mounted British military men while returning from Boston market. He reported his observations to Sergeant William Monroe of the Lexington militia, proprietor of Monroe's Tavern in Lexington, a short walk from Buckman's.

That night Monroe's Tavern was filled with drovers who, as was their custom, would later sleep on the floor of Sergeant Monroe's large upstairs meeting room. Unaware of anything out of the ordinary, the turkey drovers in particular planned to allow extra time in

Elbridge Gerry

the morning for luring their turkey flocks down from their roosts in the trees with a breakfast of grain spread on the ground before resuming their gabbling, wing-flapping stroll to market and the waiting fireplace spits of Boston's kitchens.

In the town of Menotomy, five miles to the east of Lexington, Elbridge Gerry, a prosperous merchant and an active Whig, was surprised when two fellow members of the local Provincial Committee of Safety opened the door and hastily reentered. They had left a meeting of the committee just a short time before.

They reported that while riding homeward in a chaise they had come upon eight mounted British officers who had just dined in Cambridge. It was most unusual for British military personnel to be abroad after dark anywhere outside Boston, and these eight riders had been spotted just at sunset moving away from the city. Gerry, suspecting a British strike, considered his next move.

A member of the now-prorogued Massachusetts Legislative Assembly, a delegate to the extralegal Provincial Council that had replaced the assembly, and a delegate of the illegal Marblehead Committee of Correspondence, the thirty-one-year-old Gerry was currently engaged with smuggling food donated by the other colonies through the port of Marblehead to the local residents. He was also serving on the Committee of Safety, part of the shadow government responsible for creating a network of town militias that in turn were training men, including the Minutemen, to fight the redcoats.

On top of all these other assignments, Gerry was equally busy as chairman of the Committee of Supply, which was concerned with military logistics, army supplies, and quartermastering. This was a task for which he was ideally suited, for he was a prominent businessman who headed a large firm that exported dried codfish to Barbados and Spain. Despite a reputation for truculence, he was also a skillful politician who would show his great courage the following year by signing the Declaration of Independence.

Gerry decided to send an express messenger to Lexington, five miles to the west.

In Lexington, not long before eight that evening, the Reverend Jonas Clarke received Gerry's message in his parsonage and immediately informed his two houseguests, Samuel Adams and John Hancock. They had been staying there while conducting the Provincial Council meeting in Concord. Complicating matters, Hancock's aunt Lydia Hancock and his fiancée, Dorothy Quincy, were also the guests of Parson Clarke and his wife and eight children.

Eight or nine British officers had been seen riding toward Lexington, the messenger reported, perhaps looking for them. Hancock's well-known, ostentatiously gilt carriage was outside the door of the parsonage and would be easily visible from the main road, only a few hundred yards away.

Meanwhile, at Buckman's Tavern, in the second-floor room of John Lowell, personal secretary of John Hancock, lay a huge trunk, four feet long, two feet wide, and about two and a half feet high. The trunk had been made from a hollowed-out tree trunk—hence its name—then covered with leather and studded all over with nails. Hancock had custom ordered it from a trunk maker named Bell in Boston specifically to fit on the back of Hancock's carriage for his trip to the upcoming Continental Congress in Philadelphia. The trunk was stuffed with the proceedings and papers of the just-ended

Provincial Assembly—documents incriminating enough to order most of the Whig leaders of Massachusetts to the infamous gallows on Tyburn Hill in London.

Anyone, including British officers, who might come up the tavern stairs could easily find it. With the strange activities out on the roadway, Lowell felt an urgent need to move the trunk to safer quarters but it was unfortunately too heavy for one man. He set out to find some help.

Militia sergeant William Monroe, in his tavern mulling over what young Solomon Brown had told him, was also deciding that something strange was afoot. Other witnesses reported that they had seen a band of mounted British officers and sergeants pass through Lexington that evening, heading toward Concord. Leading them was Major Mitchell of the 5th Foot. A routine mission would not have required such a high-ranking officer. Sergeant Monroe wondered if the British officers were on an errand to assassinate or capture Adams and Hancock. He walked up to the Clarke parsonage to discuss the matter with them. They had surmised the same thing.

The sergeant promptly called out a squad of eight Lexington militiamen to maintain an all-night guard around the parsonage. Leaving his busy tavern in the care of a fellow townsman, Sergeant Monroe personally remained with his squad. It was eight P.M.

Samuel Adams went to bed in the room he shared with John Hancock and lay listening to the slow ticking of the tall clock that stood near the bed. A notoriously light sleeper, he could not have slept deeply. He had the most tantalizing problem of his life to think about.

Ascetic, palsied, in rumpled clothing, a Harvard graduate who had failed in several businesses, including an inherited brewery—even failed, improbably, as a tax collector—at age fifty-one, he lived in a rickety house on Purchase Street, along the Boston waterfront. Largely indifferent to money (he would later laugh at a bribe from General Gage), Adams had discovered in his thirties that, although he had no head for business, he had a positive genius for politics. As he prepared for bed in the Clarke parsonage, he was one of the most powerful politicians in his colony, clerk of the Massachusetts Assembly, second only to Hancock, who was speaker.

He also had a motive, an obsession: to free the colonies from their royal master; in particular, to return Massachusetts to a Puritan community, without all the pomp and circumstance—all the luxury and greed and posturing—he felt was ruining the colony. Everyone would go back to wearing basic black and holding one hand on the Bible, the other on the plow. And to get it, he would use—and had used—methods most men would shun, everything from tarring and feathering his critics to stirring the waterfront rabble into riots; from wildly irresponsible and inflammatory propaganda that stole away all respect for Gage and his troops to terrorizing Loyalists and their families in their homes; from vandalizing property to forcing office holders to resign with threats of physical abuse, even to preventing the court system from functioning. In short, Adams had launched a campaign of emotional dislocation and terror that had snatched away much of Gage's authority in Boston and the surrounding countryside.

Six months before—in October 1774—at the First Continental Congress in Philadelphia's Carpenter's Hall, Sam Adams pried and squeezed, wheedled and cajoled a promise from the other delegates to the Congress that if British troops attacked the people of Massachusetts the other colonies would come to their aid. But because of their deep distrust for Adams's mobbish ways, many of the delegates warned him that the British had to be the first to shoot. If Adams's followers provoked an incident—which they already had done a number of times—or if they fired first on the British, Massachusetts would find itself facing the might of England all alone.

So Adams had returned home with a passionate mission: to goad General Gage into an attack. Then he could return to the second session of the Congress in Philadelphia and there on the floor, in front of all those other delegates, characteristically standing on his tiptoes, wave a bloody shirt, bellowing, roaring, demanding revolution!

But Gage proved a difficult bull. Weeks had turned into months, the autumn into winter, winter into early spring. Gage had remained in his lair, waiting for the closing of the port of Boston to bring Massachusetts to its knees, fully submissive to King George III.

Gage had made several minor excursions seeking weapons and had engaged in one major confrontation. The previous September 1, while Adams was on his way to Philadelphia, Gage had learned that the colonists were smuggling gunpowder into their homes from the Massachusetts Provincial Powder House, on Quarry Hill in Charlestown, just six miles northwest of Boston. Gage had unexpectedly swooped out of the city, seized 250 half barrels of powder from the powder house, and marched back home with his prize.

But the next day, due to unfounded rumors of the bombing of Boston, 4,000 Minutemen rushed into Cambridge. Within three days, 30,000 men from dozens of towns in Massachusetts and Connecticut were on the march. Gage was stunned at the reaction and began to build fortifications around Boston. With only 3,000 British regulars on hand, he realized that he was not in control of the colony. Even though the armed farmers drifted back home when they found the rumors untrue, Gage was impressed enough to ask the king for 20,000 more troops.

After the affair at the powder house, Paul Revere vowed that Gage would never take the radicals by surprise again. He created his Committee of Observers, with thirty sets of eyes inside Boston, and set them to watching Gage around the clock. And that was what they were doing at the moment.

It seemed that Gage was winning the war of wills against Sam Adams. In two weeks, the Second Continental Congress would meet in Philadelphia, where, because there had been no new British provocation, the moderates would begin to beat back the radicals and prepare a petition to the king that would have the colonies offering to kiss the rod, to seek reconciliation. This would bring Samuel Adams out of his chair, beating the air with his trembling finger. It would be an infuriating and dangerous setback to Adams's fifteen years of unremitting efforts to lead Massachusetts away from England.

If Gage didn't act soon, Adams realized, there would be little chance for a revolution. So Adams was offering a carefully set trap to entice Gage with the most sought-after bait—himself and John Hancock in Parson Clarke's easily located home. King George III had ordered Gage to capture them and ship them in chains to England to be tried for high treason. Dangling from a rope on Tyburn Hill, they would be two great trophies for the general's career advancement.

Then more bait: the illegal, forbidden Provincial Council, sitting in Concord, just a short march west of Lexington, and composed of the most influential and defiant rebels in all of Massachusetts.

Then the third bit of bait, also in Concord: the rebels' cache of weapons and ammunition—even cannon.

All General Gage had to do was march out from Boston and snatch it up. And trowel himself with buckets of glory.

Of course, there really was no bait in the trap. Adams and Hancock were going to hurry away, leaving a dusty plume in the night air behind their fleeing carriage. And the delegates to the legislative council had already packed up and scattered. As for the arms, Concord radicals—men and boys—had attended to them the day before, Sunday. They had carefully removed most of the arms, removed or buried all the cannon, and even relocated most of their barrels of flour. Musket balls by the sackful had been carried off and hidden in a nearby swamp.

Adams's chances were ebbing. Yet those British horsemen—what did they mean? Was tomorrow the day? The clock continued to tick slowly in his bedroom as he thought about Gage.

By nine o'clock in Lexington, some thirty more militia had gathered on the green near Buckman's Tavern. There, militiamen Solomon Brown, Elijah Sanderson, and Jonathan Loring volunteered to ride out to observe the actions of the British officers. It was long past darkness, which had effectively ended all plowing, and most of the colonists were either in their homes or in the taverns. The roads would be mostly empty. The three young men left on their patrol, heading west, away from Boston and toward Concord, following the trail of Major Mitchell.

Back in Boston at that same hour, the crew of the HMS *Somerset* without warning suddenly made a series of significant moves. The *Somerset* had been deliberately anchored in the estuary between Charlestown and Boston for the previous four days in order to stop radical trafficking by boat. But at this hour the crew abruptly seized all the ferries and anything else on either shore that could be floated and rowed. For the rest of the night all further crossings were forbidden to anyone.

In his Boston office, Dr. Joseph Warren, learning of the *Somerset*'s actions, decided to play his trump card. He would pay a visit to an "unimpeachable source" inside the British high command, a very dangerous move he would make only on the most urgent need. He hurried through the dark streets of Boston to a rendezvous.

"Tonight," said his informant during the secret conversation. "The march is tonight; the target is Concord."

But which way would the troops go? By water? Or by land? The choice was significant.

The land route was longer, an extended semicircular march out of Boston, down Boston Neck, and around to Lexington—a distance of twenty-one miles. If Gage elected the shorter route, by water, British sailors would row the troops across Boston's Back Bay and land them on Phipp's Farm, on Lechmere Point, putting them five miles closer to their destination. To the British foot soldier, burdened with a hundred-pound pack, five miles extra each way could be a great distance. If the troops came by water, the radicals would have less time to get ready, to rouse the countryside, to surprise their targets.

By water, Warren was told. It was no surprise to him, it made tactical sense, but it was vital confirmation. Now he could act.

Dr. Warren had three ways to warn the radicals of Gage's march. He could send an express rider down Boston Neck, around to the Concord Road, and so on to Concord. There was always a grave danger that the sentries on the neck would stop the rider—especially that night. In the previous month, the guard at the fortified sentry gate had seized over 1,300 musket balls destined for the militia.

Or he could send an express rider by boat across the Charles to Charlestown, then by fast pony straight out the Concord Road. But there was a very real danger that the crews of the *Somerset* and the *Boyne* would catch anyone crossing the Charles that night. To make matters worse, a nearly full moon was expected. In a cloudless sky.

There was a third way: by signal from Boston to Charlestown, then by a Charlestown express rider to Lexington and Concord. There was a problem here also, for signaling at night could be spotted by lookouts on the British ships or by soldiers prowling the night streets.

Dr. Warren decided to hedge his bets and use all three methods. He sent for Paul Revere and another express rider named William (Billy) Dawes.

Near the hamlet of Lincoln, which lay between Lexington and Concord, the three Lexington militiamen were searching for the British army horsemen. But the redcoats seemed to have disappeared. Then, unexpectedly, at a bend in the road, several mounted British officers suddenly sprang from deep shadows under some trees. They shouted for the militiamen to halt.

Herded into a field beside the roadway, they were closely questioned by one of the officers. The British were particularly interested in locating Hancock and Adams. After several hours of interrogation, the trio was detained in the fenced-in field. The army officers resumed their wait under the trees.

In Acton, a village several miles northwest of Concord, thirty-year-old Isaac Davis, a farmer and captain of the local minutemen, and his twenty-nine-year-old wife, Hannah, examined their children and decided that two of them had a rash that looked frighteningly like the all-too-common canker rash, a fatal childhood malady. Prayer was their only resource. With heart-aching dread, they put their children to bed for the night. In many other homes children were presenting the same terrifying symptoms.

In Boston, in the darkness of ten o'clock, small groups and knots of figures began to assemble furtively on the edge of the Boston Common, along the little-frequented shore of Back Bay. Gradually the number of shadowy figures swelled into the hundreds. They spoke in hoarse whispers, struggling to find their places. In time there would be more than 700 of them, all British troops from the nearby barracks, being urged by murmured orders to form up, stand still, be quiet.

Among them were three young lieutenants named Barker, Baker, and Gould, none of whom had ever been in action before. Also present, a late and unexpected addition, was the head of the British marines, Paul Revere's neighbor Major John Pitcairn.

Considered by his friends to be gentle and understanding, he had become quite popular with the Sons of Liberty. On duty he was a different man, though—feisty, with a peppery temper and a direct, throat-grabbing manner. He was also well-known for carrying with him his prized pair of beautifully worked and etched Spanish pistols. Pitcairn's assignment was to lead the elite British advance guard.

The troops assembling in the dark hated their barracks duty in the colonies. Under the best circumstances, army life was brutal. Paid less than a chimney sweep and not respected even in England, the British soldier often lived the life of a nonperson. Military life was so harsh—and, consequently, recruiting was so difficult—that a soldier was kept in service for life, "as long as he could stand."

Desperate unemployment in England was often a cause for enlistment. Many were without a craft or a trade and had no place else to go. Some were criminals and thieves. A few glasses of free rum would usually do the job on others. Press-gangs would supply the rest.

These men were often despised by their own officers and the colonials alike. Unwelcome in the town, reviled and always in danger of attack from bands of waterfront thugs, always the target of shouted insults when on parade, subject to disease, sullenly angry, bored, broke, they lived under the harshest of disciplines.

As a result, they were prone to being drunk as often as money would allow. Lieutenant Barker, finding that rum was abundant, ferocious, and appallingly cheap in America, noted, "A man may get drunk for a copper or two." Major Pitcairn reported that he lost seven men "killed by drinking the cursed rum of this country." He predicted that rum would "kill more of us than the Yankies will." Another soldier was given 500 lashes, enough to kill most men, for selling his musket for rum.

The British regular was so prone to desertion that General Gage once transferred a contingent of troops in New York by sea rather than march them overland for fear he would loose "half of them." The colonists exacerbated his problem by forming an underground railroad for defecting redcoats and, in reply, General Gage announced a policy of shooting deserters.

At eleven o'clock, crew members of the *Somerset* and the *Boyne* rowed the longboats down to the launching site and began loading the troops. They were going to have to row against the tide.

In Boston, newspapermen Edes and Gill had reached separate decisions. Believing that General Gage's remarkable forbearance was at an end, Edes had elected to abandon his printing press and type—indeed, his entire printing business on Queen Street—and leave the city. Gill had elected to stay with the paper, along with Edes's son, Peter.

While they continued working in the print shop, the elder Edes, praying that he had not delayed his departure too long, was fleeing down the night streets and alleys of the city, under the same nearly full moon that lit the way for the British troops scrambling into navy longboats.

Just as Edes was leaving Boston, Isaiah Thomas, the publisher of *Spy,* who had been threatened by the entire 47th Regiment, was quixotically reentering the city. Having safely stowed his family in Watertown and his press in Worcester, he was returning to hold an urgent conversation with Dr. Joseph Warren, who, after dispatching Revere and Dawes, stubbornly remained in the city.

THE
AMERICAN
Revolutionary
WAR
ALL BRAVE, HEALTHY, ABLE BODIED, AND WELL
DISPOSED YOUNG MEN,
NEIGHBOURHOOD, WHO HAVE ANY INCLINATION TO JOIN THE TROOPS,
NOW RAISING UNDER
GENERAL WASHINGTON,
FOR THE DEFENCE OF THE
LIBERTIES AND INDEPENDENCE
OF THE UNITED STATES,
TAKE NOTICE,

Musket Balls

Musket Balls such as these were used during the 1700's and 1800's. These lead balls, .69 caliber in size, were commonly used during the period from the Revolutionary War to the Civil War. The most common gun used to fire these balls was a smoothbore flintlock musket which could be fired at a rate of 2 to 4 times per minute. This gave way to the "Cap Lock" musket and the conical "Minnie" ball during the Civil War.

Distributed by © Americana Souvenirs & Gifts
Gettysburg, PA 17325
USPABM-908
Designed and Printed in the USA

Thomas needed the continued support of the radicals for his paper, but, even more important, he needed Warren's help in finding newsprint. Without it, the *Spy* was out of business. He left the city that same night, later telling his readers: "I escaped myself from Boston on the 19th of April, 1775 which will be remembered in future as the Anniversary of the Battle of Lexington!"

Of the two post riders, the twenty-nine-year-old Billy Dawes had arrived at Dr. Warren's office first. A tanner by trade, Dawes was a self-confident member of the Sons of Liberty, an excellent rider, and a dedicated intriguer. Recently he had succeeded in smuggling two cannons past Gage's guards and out of Boston. An aggressive man with no love for the British army, he had once knocked down a redcoat for insulting his wife.

Billy Dawes received the message for Adams and Hancock and left before Revere showed up. Dr. Warren wondered if Dawes would get past the sentries on Boston Neck.

Revere, when he arrived, conferred only briefly with Dr. Warren, then left to go hunt up young Newman, Colonel Pulling and Bernard at ten o'clock. He found them all waiting behind Newman's house. He told them to follow the prearranged plan of displaying lanterns from the belfry of Boston's North Church to watchers across the water in Charlestown.

The signal was simple: one lantern if by land; two if by water. So Revere told them to display two. By water.

With Bernard standing guard outside the church, Newman and Pulling went inside. Each man hung a lantern around his neck with a leather thong, then climbed nearly the equivalent of fourteen stories high inside the steeple and opened the window.

It was a clear, cool night filled with stars and a nearly full moon just rising behind them in the east. They each lit a candle, closed the covers of their lanterns, and then for several brief moments held the two lanterns over their heads, aimed at Charlestown.

Across the estuary of the Charles River, posted watchers saw the light of the two pale lanterns. In seconds, the message had been received.

Newman and Pulling descended the ladders and stairs inside the steeple, then stowed the lanterns. But as they were about to leave the church, they almost stepped in front of British troops moving through the streets. What to do?

* * *

From Charlestown, a rider was dispatched to Lexington to warn Sam Adams and John Hancock. He would be hard-pressed to evade all the roadblocks set up in the darkness along the way. He was quickly swallowed by the night.

The Charlestown watch now turned back to wait for Revere.

Meantime, Newman and Pulling, groping back through the church in total darkness, came to a window near the altar. Seeing no British troops, they climbed through the window and disappeared among the dark streets of Boston's North End. (No record tells what Bernard did.)

Sometime after ten P.M., Revere went down to the shore at the North End where he kept a boat concealed under a wharf. And there he met his two rowers, Joshua Bentley, the boat builder, and Thomas Richardson. The oars were muffled—tradition says with a borrowed petticoat or two loaned by a patriotic lass—and the two rowers set out across the river to Charlestown. Looming in front of them was the HMS *Somerset.*

Along the edge of Boston Common, the British regulars formed up as best they could in the darkness and felt their way down to the beach to board the longboats, which were well-outlined on the moonlit water.

Not far from the British boarding activity, Revere's two rowers used the partial blackout caused by the Boston skyline and the confusion of the longboats to glide past the *Somerset.* No voices shouted the alarm. Undetected, the little boat reached the shore of Charlestown.

Revere hastened into the midst of the waiting Whigs and for the first time saw the mount they had reserved for him: an excellent horse, strong, well-rested and fast; a mare named Brown Beauty.

It was just about eleven o'clock when Revere climbed into the saddle and set off for his ride to Lexington, twelve miles away. Ahead of him lay a series of roadblocks manned by British horsemen commanded by Major Mitchell. Ready to arrest anyone who might try to raise an alarm, they had already captured a number of people unfortunate enough to have been on the road that night.

Revere, a noted horseman, rushed down the dark road. Suddenly he saw a pair of British riders sweep out of the shadows ahead of him and onto the roadway, blocking his path. Revere turned his horse and fled. One of the riders galloped after him in hot pursuit.

Revere quickly realized what an excellent horse he had. She easily outdistanced her pursuer and in the protective darkness carried him on a long, curving detour to the north toward Lexington.

Riding in great haste into each town, he would bang on shutters, pound on doors, turning his horse, cantering back and forth as he shouted: "Turn out! Turn out! The Regulars are out!"

The designated post rider for each of these towns then stumbled from his bed, saddled his horse, and pounded through the night to a predetermined series of other towns in a network radiating outward from Charlestown, north and westward, waking each town's or village's contingent of militia and Minutemen. And in each village, a new express rider then galloped into the darkness to his preset network of towns—an explosion of riders, all crying the same message: "Turn out! Turn out! The Regulars are out! The Regulars are out! Pass the word! The Regulars are out!"—then galloped off into the complicit darkness to the next town, the next militia unit, the next post rider. Behind them they left ringing bells, beating drums, and roaring signal muskets.

In the town of Medford, the post rider was young Dr. Martin Herrick, who had worked that day, making house calls in Lynnfield, fifteen miles south. He hastily saddled his horse and hurried to Stoneham, then Reading, then Lynn End, emulating Revere by hammering on window shutters, banging on doors, and shouting his message.

In Lynn, Danvers, Andover, and dozens of other villages, all through the night express riders arrived shouting the same dire words—"The Regulars are out!" Militia units ran to muster on village greens armed with their muskets and powder horns, then, with more beating drums, more ringing bells, quickly marched off for the confrontation with the regulars. When Dr. Herrick finished his express route, he got a musket and joined in the march as a volunteer.

The twenty British officers and sergeants strung along the road to Concord were totally inadequate to cope with the enormous network of riders who galloped along scores of other roads from village to village all through the night.

Brown Beauty was blowing and snorting and tiring fast as she carried Revere into Lexington, but she had performed superbly. It was now the first hour of Wednesday, April 19, 1775. Revere hastened past Buckman's Tavern, with its famous red front door, calling to the thirty militiamen who were waiting in the darkness or inside the tavern: "The Regulars are out!"

Following the right-hand fork down Bedford Road, Revere traveled a few hundred yards more along the edge of the fifty-acre farm that surrounded the Clarke parsonage. Overriding the protests of Sergeant Monroe, Revere hammered on the door and called out to Adams and Hancock, inside the house.

Moments later the door opened. Adams and Hancock stood there. On the second floor, Hancock's aunt Lydia and his fiancée, Dorothy Quincy, looked out from the bedroom they were sharing. At other windows appeared the heads of Parson Clarke, his wife, and some of their eight children.

"The Regulars are out!" Revere shouted.

It was a stunning moment. Everyone had heard about the rumor from London stating that General Gage had been ordered to arrest Adams and Hancock. Now the two were obviously in real danger. It was time to pack up and leave as quickly as possible.

Paul Revere asked about the Charlestown messenger who had been sent ahead of him bearing the message of the two lamps. The man had never arrived, he was told. Most likely he had been nabbed at one of the army roadblocks. So the alarm flashed by the two lanterns in the North Church steeple had proved useless—a failed effort after all.

Nor had Billy Dawes arrived. Possibly he, too, had been captured. Had Revere been taken by the crew of the *Somerset* or by the band of mounted officers outside Charlestown, no warning would have reached Lexington. Samuel Adams and John Hancock would have been sound asleep in their bed when the regulars marched unannounced into the village. And the trunk with Hancock's papers would have sat accusingly in Buckman's Tavern.

Much to Revere's relief, Billy Dawes arrived a half hour later. On his slower horse, he had traversed seventeen miles in three hours. But by the inflexible historic rule of Second Place Is No Place, William Dawes has almost fallen out of the history books. The ride celebrated in story and poem features just Paul Revere.

Upon Revere's arrival, Sergeant Monroe and other militiamen had roused the entire militia of Lexington, who soon came running for muster on the green by Buckman's Tavern.

Now a new round of express riders galloped away from the village, so many that neither Revere nor Dawes could find a fresh horse. But their work that night was not finished. Together they were now sent to finish their mission—to spread the alarm to Concord on two very tired mounts.

For Samuel Adams, standing in the doorway of the parsonage, it was a moment of near triumph. The answer was in—yes. Gage had finally taken the bait. Tonight, just one shot from a Brown Bess musket, one bloody shirt and Adams would have his revolution.

This was also a momentous event for John Hancock. He fully expected that the Second Congress would name him commander in chief. Caparisoned in a gorgeous uniform—chosen from a trunkful of new ones that had been custom-made just for him—and accompanied by furling banners and pipes and marching feet, he would lead the way into final battle against that quiet enemy, Gage. Hancock had never been more transported: commander in chief.

Also greatly pleasing to him would be the perceived ascent he would make—a classic rags-to-riches tale, up from a childhood of near poverty, much of which had been spent in this Hancock-Clarke house.

Built frugally by his minister grandfather, the house had been completely made over when Hancock's bachelor uncle, Thomas, had become enormously rich in shipping. After his minister father died, Hancock, then six, was taken by his grandparents into the parsonage and lived there for seven years. Only years later did Hancock become heir to the uncle's fortune—one of the greatest in New England. Detractors might say that he was wealthy from another man's riches, but he would be able to claim that he had become commander in chief on his own merit.

In addition, a break with England would free him from the 500 indictments for smuggling that were pending against him in the courts.

As for his aunt, Lydia Hancock, yet to come was her moment of triumphant matchmaking. She was aunt to both Hancock and Mrs. Clarke, Hancock's first cousin, and felt herself close enough to almost be Hancock's true mother, with a mother's right to matchmake. For four years she had been trying to bring the aging bachelor of thirty-eight to the altar with the "slight and sprightly" (and increasingly, at the age of twenty-seven, spinsterish) Dorothy Quincy, of *the* socially prominent Boston Quincys—and thereby produce new male heirs for the Hancock family.

She had brought Dorothy Quincy to the Clarke parsonage when almost all of the Boston radicals—including Hancock—had fled the city earlier in the month. Her primary reason for leaving Boston was the persistent rumor that the British navy was about to bombard the city and burn it to the ground. But it also served her plans to have Dorothy under the same roof as her elusive nephew.

An exultant, expectant Sam Adams, now fully dressed, and always tending to business, walked the few hundred yards down Bedford Road with Hancock and Reverend Clarke to Buckman's Tavern, on the Commons, to chat with the armed militiamen. In light of the events that were about to take place on the Lexington green, what he would say to them has been the subject of debate among historians concerning the firing of the first shot.

Tall, tubercular, Captain John Parker had gotten out of a sickbed and walked the two miles to the green, where he now looked over the situation with weary, feverish eyes. A former ranger from the French and Indian War, he got his fellow townsmen to fall out on the green in orderly military fashion.

Having assembled more than three hours before, the thirty Lexington militiamen had grown weary with waiting, but now, with Adams and Hancock and the town parson observing, everyone, in the customary manner of American militia, discussed the approach of the redcoats—thirty farmers against 700 trained soldiers. They decided to send two post riders east to get more information about the regulars and their probable time of arrival.

Paul Revere and Billy Dawes, moving toward Concord, met a fashionably dressed young man on a fine mount. Dr. Samuel Prescott was on his way home from a visit with Miss Mulliken of Lexington, who had agreed to marry him. He was also a dedicated Son of Liberty, brother to one of Revere's express riders. He eagerly offered to help them spread the alarm. Stopping at every farmhouse, they roused the residents, who were then urged to spread the alarm to their neighbors. The three riders were unaware of the trap they were riding into.

* * *

Major Edward Mitchell of the 5th Foot and the members of his mounted patrol were weary, hungry, saddle-sore, and short-tempered. Mitchell had led them out of Boston at eight o'clock the previous morning. He had strung them out in small bunches over a number of miles. And now, wearily, irritably, he waited for further orders, not knowing what was going on, how much longer his patrol would be here, or when it would be relieved or by whom.

He had remained on the road to Concord, waiting to catch any messengers who might be abroad. So far, in the fenced pasture behind them, he and his group had four men to show for sixteen hours' labor, including the three Lexington militiamen.

Mitchell's horse pricked up his ears. Riders were approaching. Ready for trouble and not counting on relief, he and three of his officers stepped out on the roadway waving pistols and swords and shouting threats. They bagged a bonanza: three more riders. With curses and additional threats, they forced the three men to turn into their detention pasture.

Suddenly, apparently at a prearranged signal, the three bolted. One, Dr. Samuel Prescott, who knew the neighborhood thoroughly and was well mounted on a fresh horse, jumped a stone wall, evaded his pursuers, and quickly disappeared.

Paul Revere galloped to the right and ran into a group of six more British horsemen. Trapped. Now with ten horsemen surrounding him, he watched as Dawes bolted. Dawes had almost escaped when his horse shied, threw him, then galloped away in the moonlight. Dawes quickly crawled into the brush to make good his escape, though, and so groped his way back to Lexington on foot and into anonymity.

Holding a pistol against Revere's head, Major Mitchell began an interrogation. The answers he received stunned him.

First, the prisoner shocked Mitchell by admitting that he was indeed the notorious radical Paul Revere. He then informed Mitchell that the British regulars were on the move and gave more details about the march than Mitchell had. He also warned Mitchell that the army was marching into a trap because the militia throughout the whole countryside had been alerted. Soon, he warned, more than 500 militia would muster in Lexington alone. Since Mitchell and his men had heard the sounds of alarm shots and warning church bells clanging in the night, they were convinced.

Mitchell decided to take his troops and ride back toward Boston to warn the main body of marching soldiers. He ordered a sergeant who was poorly mounted to take Revere's horse, Brown Beauty. Leaving Revere on foot with the four others who had been detained, the soldiers cantered away, Revere, thinking his work for the night was done, walked back to Lexington.

But when he reached town, he was astonished to find that Hancock and Adams hadn't fled. Milling about in the parsonage, they were still arguing, hours after he had first warned them and only a short distance from the main road where mounted army officers were passing up and down.

In the background, Hancock's aunt, in "high hysterics," Dorothy Quincy, and the large Clarke family contributed to the confusion as they tried to pack, dress the children, and hide the family's meager possessions from the notorious looting of the British regulars.

In the midst of it all, John Hancock was sharpening his sword and vowing that if he had his musket with him he would stand shoulder to shoulder with the troops assembled on the town green. Apparently, no one suggested to him that he could borrow a musket for that purpose.

Sam Adams had a simpler agenda. He wanted to leave. Immediately.

Revere sided with Adams. Hancock and Adams were the two most important, most powerful men among the radicals of Massachusetts, and they had vital work ahead of them in Philadelphia. He urged them to set out immediately for a place of safety.

On the green, Captain Parker and his militia listened for the dull clump of British boot soles approaching in the darkness. Yet still there was only silence. The tense stand continued.

Finally, Hancock's carriage went rattling down back roads, its gilt and rococo decoration incongruous in the rude April countryside. The departure had been so abrupt that Adams left behind the entire wardrobe he'd intended to carry to Philadelphia. Bearing Hancock, Adams, Revere, Sergeant Monroe, and Hancock's personal secretary, John Lowell, the carriage headed for the home of a clergyman's widow in Woburn, a few miles north.

Hancock now learned from Lowell that the vitally important trunk had never been strapped on the rack of the carriage where it belonged but had been left in Lowell's room. Revere and Lowell agreed to go back to retrieve the trunk and bring it to Hancock in Woburn in a wagon.

When he reached Woburn, Hancock remembered something else—a salmon he had been given was still in the kitchen of the Clarke parsonage. So he sent back his coach to pick up the fish. Strangely, while fetching it the carriage did not retrieve the much more important trunk or Adams's much needed wardrobe. He had only the clothes on his back to last for the next weeks and months in Philadelphia. The lateness of the hour notwithstanding, when the coach returned, Hancock asked that the fish be cooked immediately. Excitement always made him ravenous.

In Lexington, Paul Revere found that in his absence one of the two mounted scouts Captain Parker had dispatched had returned to announce that there was no army on the road. He had found nothing. The whole military maneuver was a feint of no consequence. Just another false alarm.

But as Revere was entering the tavern to get Hancock's trunk, the other mounted scout came pounding up the roadway yelling that the regulars were right behind him, little more than a mile away and doing a ground-eating quick-march.

Captain Parker ordered the drummer to beat a call to arms. Then he assembled his militia unit—now seventy-seven strong—and lined it up in two ranks along the green. He was an experienced officer with combat experience who understood that the regulars would immediately take this posture as a challenge. That he meant to do battle was indicated in his instructions to the troops: "Don't fire unless fired upon. But if they want a war let it begin here."

The militiamen stood in silence in the gathering light of dawn. The stirring birds called in the trees, and the odor of fresh-turned earth and apple blossoms hung in the air. Small knots of onlookers stood about on the common. Women and children peered from the windows of the houses around the green. Dorothy Quincy continued her vigil in the second-floor bedroom of the Clarke house.

All listened for the sound of British boot soles. It was five A.M.

At the first faint light at the tavern windows, Revere and Lowell stepped out of Buckman's and into the midst of the crowd of militia and onlookers, struggling with the immensely heavy trunk, its brass fittings and the nail studs on the leather covering glimmering.

In Woburn, meantime, Hancock had just tucked up his napkin

to feast on the salmon when a messenger rushed in to say that the regulars were on their way. Hancock and Adams, in a panic, ordered that the highly visible Hancock coach be put in some trees while they concealed themselves in the woods. The salmon was left on the table, uneaten.

They waited, but no regulars appeared. Tentatively, they emerged from hiding and decided to move farther away to yet another "modest" house—that of Amos Wyman.

In Lexington, with shocking suddenness, the advance army unit under command of Marine Major John Pitcairn abruptly hove into sight on the roadway, quick-marching directly toward the green and the militia assembled there.

With Pitcairn at the head of the column were three other officers, including Major Edward Mitchell, who must have been saddle-sore in the extreme, having been mounted, and without any sleep, for nearly twenty-four hours. On foot were Lieutenant Barker and Lieutenant Gould, who would have prominent roles in the action that followed that day.

From his horse, Major Pitcairn cried out to the militia: "Throw down your arms! Ye villians, ye rebels."

Colonel Parker ordered his men to disperse, and they started to obey. None had thrown down their arms, but many of those who had heard Parker's command turned away and showed their departing backs to the British.

Then a shot was fired.

Long months of pent-up resentment in the British troops exploded. Apparently without orders from any of their officers, the regulars fired a volley at the militiamen, then charged with their bayonets. Their officers were unable to control them.

At the height of this action, Colonel Francis Smith arrived with the main contingent of troops and saw British regulars running amok, firing their guns, preparing to assault private houses, and moving menacingly toward Buckman's Tavern. Smith turned to a drummer and ordered him to play "Down Arms" over and over. Finally, reluctantly, the troops came to a heel. When the officers restored the angry troops to their ranks, the town green revealed a shocking scene.

Militia Captain Parker's kinsman, Jonas Parker had been wounded, and while lying on the ground attempting to load his gun, he had then been bayoneted fatally. Townsman Jonathan Harrington fell, blood flowing from his breast. With his wife watching from a window in their house, he stood up and stretched out his hands toward her, then fell again. On his hands and knees, he crawled across the road. She ran to the front door, opened it, and looked down. There, on their doorstep, he died. Two others running from the common were shot, presumably in the back.

In all, eight militiamen lay dead and nine others were wounded. If Colonel Smith's arrival had been delayed a few more minutes, the carnage might have been much greater.

Only a few of the militia had managed to return fire. Jonas Parker, John Monroe, Ebenezer Monroe, Jr., and others got off shots before leaving the line. Solomon Brown and Jonas Brown fired from behind a stone wall; one other person fired from the back door of Buckman's house; Nathan Monroe, Lieutenant Benjamin Tidd, and others retreated a short distance, turned, and fired.

Major Pitcairn's horse had received two minor injuries, and one regular, Private Johnson on the 10th Foot, was slightly wounded in the thigh. Johnson's luck didn't hold for long; he was later wounded mortally at Bunker Hill.

There were many eyewitnesses to the action at Lexington: militia; British regulars, officers and troops; and bystanders; as well as the women and children looking out from the windows of the houses around the common. Many kept diaries, and still more testified in affidavits collected later by the Committee of Safety.

But even today the question persists: Who fired the first shot?

One of the British officers who stood on Lexington Green and saw the flash of that first shot was Lieutenant John Barker of the King's Own Regiment. He asserted that the militia fired first.

"On our coming near them," he said, "they fired one or two shots, upon which our men without any orders rushed in upon them, fired and put 'em to flight. Several of them were killed, we could not tell how many because they were got behind walls and into the woods."

But few others agree with that view.

"Half a gunshot" away from the regulars, Revere and Lowell, dangerously exposed and feeling musket balls whiz past their heads, witnessed the whole action: "While we were getting the trunk, we saw the regulars very near, upon a full march. We hurried towards Mr. Clarke's house. In our way we passed through the militia. There were about fifty. When we had got about one hundred yards from the meeting-house, the British troops appeared on both sides of the meeting-house. In their front was an officer on horseback. They made a short halt when I saw, and heard, a gun fired, which appeared to be a pistol. Then I could distinguish two guns, and then a continual roar of musketry; when we made off with the trunk." They had come dangerously close to being wounded or killed.

Professor Fischer states: "Nearly everyone, British and American, agreed that the first shot did not come from the ranks of Captain Parker's militia, or from the rank and file of the British infantry." Many militiamen contended that the first shot did not come from the foot soldiers but from one of the British officers, probably mounted, probably from a pistol.

However, the British officers unanimously believed that the first shot came from behind a hedge or a stone wall. Major Pitcairn was absolutely convinced that four or five shots were fired from over a wall, and other British regulars believed the first shot came from Buckman's Tavern.

Professor Fischer has concluded after extensive study that it is probable that several shots were fired "close together—one by a mounted British officer, and another by an American spectator." He adds that one of the shots might have been fired deliberately "either from an emotion of the moment, or a cold-blooded intention to create an incident. More likely, there was an accident."

"Cold-blooded intention to create an incident"—does he mean Sam Adams?

The British military historian Brendan Morrisey takes to task Captain Parker, a combat veteran who should have known better, for lining up "his men in a suicidal, and militarily useless, position that morning." Concerning that first shot, Morrisey asserts: "The finger of suspicion points most strongly at someone acting on orders from Samuel Adams."

Not provable. But quite plausible.

After fifteen years of confronting royal governors, customs collectors, Loyalist judges, colonial secretaries, generals, admirals, and the king himself, Sam Adams had become predictable. Always working behind the scenes, Adams craved anonymity. He never appeared in any of his political theater productions, but he was endowed with a great sense of the dramatic. He was the master American playwright of the eighteenth century, but never an actor.

Sam Adams was not the type of man who would leave anything to chance. With those British troops on the road—and American militiamen waiting for them, with orders not to fire first—Adams had found himself in a classic situation. All he would have to do was light the fuse. With the two antagonists facing each other, he could arrange to have someone—not himself; he would be far away—fire a single shot from behind a wall or a window that would provoke the British troops into a violent response against the militia. Afterward the British regulars would be accused of having fired first.

For the price of a few dead farmers, Adams could buy his war.

Colonel Smith kept his men standing in the growing light while he gradually restored order. He was probably not the best choice for this assignment. Enormously heavy and slow, he was called lazy by some, a time server by others. In any event, he had been personally chosen for this operation by General Gage himself. An "undistinguished officer, earning consistently poor performance ratings," he would be "promoted eventually to high rank, though only in home commands."

Colonel Smith now told his officers for the first time what their destination was—Concord. They were to find and destroy a cache of arms.

They were stunned. Their troops had run amok and had caused a thunderingly violent incident. A far less emotional incident during the previous September—a bloodless raid on a powder magazine in Charlestown—had caused 20,000 militiamen to swarm around Boston. Surely, many officers felt, with the countryside being aroused against them once again, the prudent strategy would be to withdraw. Seven hundred men would be no match for thousands of armed farmers, no matter how inept. Colonel Smith, however, insisted on carrying out his orders.

Major Pitcairn agreed with Smith. He had little respect for these armed civilians. Just the month before, Pitcairn had said, "I am satisfied that one active campaign, a smart action, and burning two or three of their towns, will set everything to rights."

He was basing this on the behavior of the Americans during his military expedition to Salem in March. The British soldiers had been told to expect trouble but they found none. Yet after they left, Major Pitcairn recalled contemptuously, "the people got arms and paraded about, and swore if he had stayed half an hour longer they would have cut him to pieces.''

While local residents were standing over the dead and wounded, Colonel Smith permitted the entire unit to fire one volley in the air in victory salute and give three cheers. Accompanied by beating drums, the army marched out of Lexington on the road to Concord.

Captain Parker and his militia stared in shock at the departing lobsterbacks. Two emotions were growing among the men: grief and the desire for revenge. The British had not seen the last of the Lexington contingent.

Nor did Colonel Smith have any idea what lay ahead of him.

On the road to the Wyman house, Hancock, regretting the loss of his salmon, and Adams were bouncing along in their carriage when Adams heard musket fire. It was coming from the direction of Lexington.

He applauded.

''This is a glorious day,'' he said to Hancock.

''Yes, yes.''

''For America, I mean.''

It was now dawn, April 19, 1775, and he had his revolution.

Meantime, without his salmon, Hancock had to settle for a hard-scrabble farm breakfast of biscuits and fatback.

After Dr. Prescott bolted over the wall and escaped capture by Major Mitchell and his mounted patrol, he traveled rapidly toward Concord, waking people and express riders all the way with his cry: ''The Regulars are out!'' By two o'clock in the morning, the town bell of Lincoln was ringing.

By then Dr. Prescott had already arrived in Concord and roused the militia leaders there, including Reverend William Emerson. By two A.M., Concord's bell was also pealing.

Paul Revere's intelligence-gathering system had worked well: Prescott was able to tell the town leaders the approximate number of troops on the march—800; how they traveled—by boat across the river to Cambridge; their destination—Concord; and their purpose—to capture arms.

As the militia leaders conferred, other express riders, including Dr. Prescott's physician brother, Abel, were hurrying the news to Sudbury and Framingham, Natick and Needham, Dover Farms and beyond, traveling ever farther west into Massachusetts and even curving back toward Boston.

Reverend William Emerson stood in the midst of the beautiful April morning when the farmers should have been out in the fields with their plows and oxen, and shook his head.

"Let us stand our ground," he said. "If we die, let us die here."

"Here" was an old farming town lying eighteen miles west of Boston. It had been named Concord when it was founded in 1635, not long after Boston itself was founded. Reverend Emerson and the other town leaders could look out at the surrounding valley, which stood at the confluence of three small rivers—the Sudbury and the Assabet, which form the Concord. The town was surrounded by low hills and well-established bottomland farms that had been handed down within families for generations. To serve the farms on the other side of the rivers to the west, the township had built two sturdy plank-and-log bridges wide enough to accommodate farm carts, the South Bridge and the North Bridge. The latter would accommodate a great deal more by day's end.

In Acton, before dawn, Minuteman Captain Isaac Davis looked at the armed men who had pressed into his kitchen, then started toward the door and the march to Concord. He paused, turned, and looked at his wife, Hannah, who had their sick and crying children to deal with. "Take care of the children," he said. Then he left.

Hannah Davis had a premonition that she would never see her husband alive again.

General Gage's information about the military supplies in Concord had been absolutely correct. On April 16, just three days before the attack, John Adams's cohort, the brilliant physician James Warren, wrote to his wife, Mercy: "All things wear a warlike appearance here. This Town is full of Cannon, ammunition, stores, etc., and the Army long for them and they want nothing but strength to Induce an attempt on them. The people are ready and determine[d] to defend this country inch by inch."

But by Monday the bulk of the arms had been moved to other towns such as Acton and Worcester. And while Colonel Smith was leading his band of 700 British regulars toward Concord, militia colonel and farmer James Barrett and his sons were urgently trying to conceal the remaining weapons. As the redcoats drew closer, Barrett's sons hastily plowed a field, lay the weapons in the furrows, and covered them over.

Appareled in handsome red or blue coats and their variegated regimental garb and headgear, holding proudly their fluttering regimental colors, the British regulars, with Colonel Smith riding near the lead, made a smart approach to Concord. Their beating drums and rhythmic steps were calculated to intimidate anyone foolish enough to attack. With a history of daring deeds, bravery, and great victories, the British regulars, as put upon as each soldier was, were together the terror of Europe.

The militiamen watched as the highly disciplined column approached the town "in perfect cadence, intimidating and irresistible, their arms glittering in the early sunshine." The locals, greatly outnumbered, knew that armed contingents from other towns were on the way. So they decided to occupy the high ground across North Bridge on Punkatasset Hill and wait for reinforcements.

They were watching closely when the army column halted below them. In full view, Colonel Smith held a brief conference with his officers—including Major Pitcairn, Lieutenant Gould, and a disapproving Lieutenant Barker, who was increasingly disgusted with Smith's lackadaisical command. Colonel Smith issued his instructions. It was shortly after seven A.M.

The militia observed Colonel Smith send seven companies in two contingents through the village and across the North Bridge—one group to the Buttrick farm and another several miles beyond that to the Barrett farm; Tory informers had identified both as large arms depositories. Other regulars searched the town itself while Lieutenant Gould, under Captain Laurie, led his unit of three companies to a post overlooking the North Bridge.

Never having been in battle before, Lieutenant Gould regarded the increasing number of militia staring down on him from the hill across the river with growing apprehension. They were only 800 yards apart. And the militia held the high ground.

Guided by Loyalist intelligence from Boston, Major Pitcairn believed that Ephraim Jones, owner of the Jones Inn, town jailer, and fervent radical, had secreted three cannon somewhere in Concord and its environs. After he had helped dispatch the troops on their various errands he personally went in search of those cannon. First he had to find Jones, which he did readily enough. Pitcairn banged on the door of the inn. Jones refused to open it. Pitcairn summoned some grenadiers, who broke it down.

Pitcairn's temper, always volatile, was still erupting from the riot of the regulars in Lexington. He eyed Jones, who looked back just as defiantly, and decided on the direct approach. He knocked Jones down. Then he placed the barrel of one of his Spanish pistols against the man's head, cocked it loudly, and ordered the innkeeper to reveal the hiding place of the three twenty-four-pound cannon. Jones needed only to look up into the eyes of the outraged Pitcairn to decide that the marine was deadly serious.

He led Pitcairn out of the inn and into the prison yard next door. And there they sat—three twenty-four-pound cannon along with all accoutrements, sponges, rammers, and the rest. Because of their size, the town had been unable to hide them.

In the jail, Pitcairn found two prisoners, one a Tory. Pitcairn released the Tory. He then led Jones back to the inn and ordered breakfast. He ate it and paid the bill. Meantime, Jones did a brisk business selling rum to the regulars.

Reverend Emerson mingled with the frightened people of the town, giving particular encouragement and comfort to the terrified women and children, as the brusque British regulars searched every home. The residents had good reason to be upset—Smith's men were busy looting. The nimble-fingered regulars made off with saddles, bridles, stirrups, the meetinghouse Bible, other books, shirts, shoes, pewter plates, silver buttons—even a glass of salt—and much more, the bulk of it from people who could ill afford the loss.

Smith's junior officers were watching anxiously the growing number of colonial militia. Companies from Acton, Bedford, Lincoln, and other surrounding towns now brought the count up to 500 men. Then more, then more again. Soon there were two full regiments. Colonel Smith seemed indifferent to his increasing danger.

Captain Laurie, with fewer than 100 men, and operating far beyond the North Bridge, sent an urgent message to Colonel Smith for reinforcements. Smith ignored the request while he continued stubbornly frisking the village's cupboards, barns, and jakes. His troops even broke open sixty barrels of flour but, to their chagrin, found nothing but flour.

Finally, Smith had to recognize that, except for the three cannon, he had found the roost largely empty, the chickens and the eggs long gone. But for the cannon, all he had to show for his work were 500 pounds of lead musket balls, which he dumped into the town pond and which were later recovered by the townspeople, a few wooden gun carriages, and a few barrels of wooden spoons and trenchers.

The frustrated Smith, realizing that he was going to depart almost empty-handed, decided to burn what little he had found. He ordered his men to build a fire composed primarily of the wooden spoons and the gun carriages. But the fire was too close to the courthouse, and soon shooting flames and sparks set that building on fire.

Mrs. Moulton, an aging resident of the town, with considerable courage and towering indignation, demanded that Smith put out the fire. Having second thoughts about burning the town, Smith nodded his head, and his troops aided the residents in smothering the blaze. There was little damage, and Smith had to satisfy himself with a lame insult: He had his troops cut down the town's liberty pole. Then he waited for his search units to return from the Buttrick and Barrett farms.

The column of smoke rising from the town, however, had alarmed the militia up on Punkatasset Hill. They believed the redcoats had put their town to the torch. After a hasty conference, Colonel Barrett ordered his men to form a line and face the bridge. Then he ordered them to load their weapons. Between the militia and the town, guarding the North Bridge, was the contingent that included Lieutenant Gould's company. The militia would have to get past these troops to enter the town.

Barrett now ordered the militia to march in two rows toward the bridge. Captain Isaac Davis and his Acton Minutemen led the way because they were fully equipped with muskets, bayonets, and cartridge boxes.

Lieutenant Gould and his regulars, outnumbered four to one, withdrew across the North Bridge. There the troops were ordered to adopt a street-fighting defensive position they had not trained for. While they struggled to form up, a musket went off. Without orders, a regular had fired at the colonists. Several more fired, and the army unit commenced an irregular volley. Once again, as in Lexington, British discipline had broken down.

One of the militiamen hit was Captain Isaac Davis. A British ball passed through his heart and proved his wife's premonition true. Several others were wounded, but the militia did not falter. They approached the bridge with far more discipline than the army regulars were showing. And now it was their turn.

Facing down the regulars' scattered volley, Colonel Buttrick put his men in firing position and gave the order: "Fire, fellow soldiers; for God's sake, fire!"

At fifty yards, the militiamen, each with recent weeks of military training and years of practice hunting small game, were deadly. Four of the eight army officers fell in the first volley, along with five regulars. Abruptly the army unit broke and, ignoring the bellowed commands of their remaining officers, ran back to Concord with the wounded limping after them.

At that moment a vital transformation took place among the militia. They discovered they could cow the vaunted British redcoats.

One of the wounded officers was Lieutenant Edward Thorton Gould of His Majesty's Own Regiment of Foot. He was hit, ironically, in the foot, and thus began one of the most unusual odysseys of the entire, battle-filled day.

While Gould struggled with his wound, a twenty-one-year-old Concord man, Ammi White, arrived, carrying a hatchet, to join the militia. Seeing a fallen soldier lying in pain from his wound, White brained him with his hatchet, partly scalping him and laying open his skull. Then he left the man there by the bridge to slowly die.

Colonel Smith, realizing that the militia had come between his two search parties on the farms and the North Bridge, now decided personally to protect the bridge. But Colonel Buttrick had sent half his militia force over the bridge toward Concord with orders to take a defensive position behind a stone wall up on a hill with orders to fire only on command. They waited for Colonel Smith to arrive. But this "fat heavy man," as Lieutenant Barker contemptuously described Colonel Smith, was very slow in his actions. And when he did arrive, Smith halted out of musket range and studied the militia position. Then he withdrew his forces.

Luck was with him. The four units that had been searching farms beyond the North Bridge heard the shooting and were now rushing back. They found the bridge under siege and wounded and dead regulars on the ground. Fearful of being cut off, they ran across the bridge, straight past militiamen on both sides of the river. Not a shot was fired at them.

But when the troops hurried past the still-living regular whose head had been cut open with the hatchet, they were shocked by the atrocity. Ensign Jeremy Lister would later claim that *four* men had been "scalped, their Eyes goug'd, their Noses and Ears cut off; such barbarity exercis'd upon the Corps could scarcely be paralleled by the most uncivilized Savages." Ammi White's handiwork would set off a series of atrocities on both sides during the rest of the day and create a scandal in Europe.

Meantime, in Watertown, some ten miles west of Boston, where the Committee of Safety was setting up a provisional government, publisher Benjamin Edes had located an old press. He was now busy blowing the dust off it, checking the type boxes, scrounging a supply of ink and very little paper, and getting ready for publication. The scorching verbal assault on King George III and his empire was about to continue, goring the royal bull without mercy. Colonel Smith and his regulars, searching Concord's pantries, were busy providing abundant copy.

By noon, Colonel Smith had organized his contingent for departure. The whole operation had occupied four hours. Smith put his wounded officers, including Lieutenant Gould, into commandeered horse-drawn chaises, let the walking wounded fend for themselves, and, following the military practice of the day, left the most severely wounded in Concord—in the care of the angry locals, who would nonetheless provide medical attention. By breaking their mounting trunnions—essential for swiveling their barrels—he rendered the three cannon useless, then turned back toward Boston and home and marched out of Concord in cadence.

What he had found in Concord was hardly worth one musket ball. But what his troops had pilfered was never forgotten, and what he left behind was a village filled with anger.

There would be one more score to settle as well. The younger brother of Dr. Samuel Prescott, Dr. Abel Prescott, having returned from a night of express riding to Sudbury and Framingham, was watching the army's departure when he was shot. He would die the following August.

More militia from other towns now arrived and heard the stories of the British regulars' behavior in Concord and the "massacre" in

Lexington. Colonel Smith could now look up and see above him large numbers of newly arrived militia ranging themselves along the hilltops. Some of those units had walked all night, and as they watched him they loaded their muskets and waited. They hadn't walked twenty miles just to stand and stare. They had come to fight.

Colonel Smith faced a gauntlet eighteen miles long. Most of his troops had never been in battle before and had already exhibited the unfleshed recruit's tendency to aim too high and miss his target. Furthermore, they had come with only the standard thirty-six balls and powder and no reserves of ammunition.

Smith sent flankers out on both sides of the road. Their assignment was to keep the militia out of shooting range. The flankers began by clearing one hill followed by a series of meadows. But then came Merriam's Corner. Here the road crossed a narrow bridge, forcing the flankers to come down from the hill and walk along the stream toward the crossing at the bridge. That let the militia move to within shooting range of the main body. And now they outnumbered the regulars by more than 1,000 men.

The first shot was fired after a militiaman took aim at the mass of army uniforms. Abruptly, a rain of lead balls fell on the bunched-up soldiers. There at Merriam's Corner the running battle to Boston had begun. Smith's forces were going to have to fight their way through one ambush after another, sometimes getting caught in a raking cross fire, all the time moving ever closer to panic as the casualties mounted and the ambuscades increased.

An east wind blew the dense clouds of gun smoke into the faces of the combatants. The acrid choking odor of gunpowder was everywhere, adding to the terror. Colonel Smith had no way of knowing that his unit was stacked against a frightening number of militia—and that the number kept increasing as more and more contingents from all the towns along the way joined in. Soon there were more than 2,000 militia firing at the red and blue coats.

Thousands of men arrived from Tewksbury and Reading, Framingham and Sudbury, Bedford and Chelmsford and Belerica, Woburn and Dracot and Stow and Westford, Cambridge, Arlington, Brookline, Danvers—the list was growing by the hour. And as his army stumbled along, growing more tired, wasting ammunition, at every bend in the road Smith was encountering fresh militia with fresh eyes and legs, attacking ferociously, and as he made his way homeward there were still more militia units from dozens of other towns taking up their positions and waiting for him.

Militia Lieutenant Joseph Hayward, an old Indian fighter, made it his business to pursue the chaises carrying the wounded British officers. When the sixty-year-old Hayward caught up with them, he shot two where they sat, tumbled their bodies onto the roadway, and led the chaises back to Concord. Inside another chaise, Lieutenant Gould, in pain from his foot wound, managed to keep his horse moving at a regular clip, amazingly surviving one ambuscade after another. With the whistling of musket balls filling the air, and British soldiers falling all around him, he must have wondered how long his luck would hold out.

Under the intense fire, Major Pitcairn's horse bolted, throwing its rider. Later the horse was led back to Concord and the saddlebags were searched. Inside were found the major's prized Spanish pistols.

Waiting for the regulars were the tall and consumptive Captain John Parker and his smoldering Lexington militia, some of them wearing bandages on wounds earlier inflicted by the British. Panting with hatred and looking for the signal from Parker, they let the regulars come within range of their muskets. And still they held their fire.

Then Parker gave the order. The Lexington militia rose, aimed, and fired, quickly reloaded, and fired again. Redcoats were riddled and fell across the roadway under a storm of fire so heavy it stunned the regulars into a momentary standstill. Among the casualties was Colonel Francis Smith himself, shot from his saddle with a severe wound in his thigh. Presenting such a large target high upon his tired horse, he was fortunate to have gotten that far in one piece.

Flankers finally drove off the Lexington militia.

Beyond this point more punishment was waiting from two ambuscades at places called the Bluff and Fiske's Hill. Despite the best efforts of the flankers, the militia, now in huge numbers, lined both sides of the road and fired at will.

The militia was suddenly confronted with a surprising problem: prisoners, some in desperate condition. Militiaman Simonds of Lexington took a captive who turned out to be a fifer, little more than a child, whose closely buttoned jacket concealed a fatal wound. Despite attentive nursing, the boy died a few days later.

The army's condition worsened. The soldiers were exhausted, they were running out of ammunition, and they were so thirsty—as only men in battle can be—that some of the most savage fighting occurred around wells and brooks. The toll of officers was so great that sergeants were now filling command roles.

As the army moved, it left strewn behind it amid the dead and wounded on the road a great miscellany of military equipment, abandoned muskets, hats, bayonets, cartridge boxes, leggings, coats, vests, straps, fifes, drums, dead horses, and even some of the loot stolen from Concord bureau drawers. The soldiers dropped everything and flung themselves headlong toward Boston, toward safety. But it seemed just a matter of time before all the regulars would be casualties or captives. The retreat hadn't even reached the town of Lexington.

The officers, using bayonets and threats, struggled to maintain some kind of order. They were thinking that surrender would soon be their only option. Ironically, the place of surrender might be the Lexington Common, where they had attacked the militia less than twelve hours before.

When the troops at last approached Lexington, ready to fall down and surrender, with Boston still fifteen miles away, suddenly, like a vision, they saw drawn up in line of battle and armed with cannon an entire brigade of fresh British troops under the command of Brigadier the Right Honourable Hugh Earl Percy. The exhausted troops, shouting with joy, ran into the arms of their rescuers.

Percy, commander of the British 5th Regiment, at first did not believe what he saw: a proud British task force in full flight and decimated. Percy promptly placed a cannon on each of the two hills overlooking Lexington, one on each side of the roadway, to protect the flanks of the retreating redcoats. It was an effective tactic. The pursuing militia had never faced a cannon before, and the first shot stopped them completely.

Nonetheless, Percy knew he was in a very tight spot.

He had set off from Boston at nine o'clock that morning with his troops in high spirits. The band played "Yankee Doodle," the boot soles clomped, and the wooden wheels of two cannon rumbled and clattered on the stone roadways of Boston. They were expecting just another routine excursion into the countryside.

Percy's first hint of trouble came from a rumor he heard in Menotomy at around one in the afternoon. Then more substantial news arrived. Toward him from Lexington came a horse-drawn chaise whose wounded driver had somehow managed to navigate the entire series of punishing gauntlets. Lieutenant Gould told the shocked Percy about the events he had been part of. Then he was sent on his way back to Boston in his borrowed chaise.

He wouldn't get much farther. Captured by militia in Menotomy, Gould was sent north to the town of Medford, where his wound was cared for and from whence he would give a much-publicized deposition about who had fired first at the North Bridge.

Back in Boston, rumors of the battle of Lexington began to circulate. Dr. Joseph Warren heard the first partial news at about nine o'clock in the morning and prepared to skip out of Boston immediately. The other few remaining radicals made similar plans. Those not quite quick enough faced arrest.

Percy set up his temporary headquarters in the tavern belonging to the militia's Sergeant William Monroe. There he permitted Colonel Smith's exhausted forces to rest and restore themselves while he held the provincials at bay with his cannon. He also set fire to three houses that could provide cover for provincial snipers—one of them the home in which Lydia Mulliken had become betrothed to Dr. Samuel Prescott.

Lord Percy had one potentially disastrous problem. He had not brought enough ammunition. His troops carried the same number of rounds as Smith's unit had earlier—thirty-six per man, which was hardly enough for the running battle that loomed. It also meant that he didn't have spare ammunition to rearm Smith's men. Furthermore, he had decided not to bring a carriage with a large supply of cannon balls for his two six-pound cannon, not having seen any need for it and feeling it would slow him down. So his two cannon had only the balls that were in the two small boxes on either side of their barrels. How long could he make them last?

More ominously, he was astonished to learn that Colonel Smith had not been fighting a bunch of country bumpkins firing at random from behind trees. The Massachusetts militia had been training for months under the tutelage of men with broad combat experience in the French and Indian War. These bumpkins were fighting as units, working in concert, following orders, and attacking on cue. And their numbers were increasing. Percy was greatly outnumbered.

Worst of all, Lord Percy was about to face a very gifted militia officer who had never seen a day's combat but had an exceptional tactical mind and a brilliant idea for fighting the unsuspecting regulars.

Brigadier General William Heath was now on the scene and in command of all the militia. He had learned of the movement of the regulars to Concord that morning at dawn in his farm home in Roxbury and of the fighting at Lexington somewhat afterward. Later, heading for a meeting of radicals in Watertown, he encountered his comrade, Dr. Joseph Warren, who had slipped out of Boston to go where the fighting was. Warren had a ferocious, heedless enthusiasm that was to keep the militia stirred up for the entire rest of the day. Both men were now on the scene and eager for the fight.

General Heath, a very heavy, jovial country squire, had a passionate interest in military tactics. Widely liked and respected, Heath had learned much of what he knew from books garnered from Henry Knox's London Bookstore in Boston, a favorite hangout for British officers, and from serving, such as it was, in the Ancient and Honorable Artillery Company of Boston.

Heath had spent years studying and musing over what he saw as an inevitable battle between the colonists and the British army. He had warned his countrymen about the coming violence in numerous articles in the local journals. Many had listened to him. More than just a passive theorist, Heath devised a plan of attack for his specialty—skirmishing. Amateur or not, Heath had information about skirmishing that was probably matched by only a few military specialists on either side of the ocean. And Lord Percy presented him with the perfect laboratory to put his theories into practice.

His plan was to put a moving ring of fire around the marching British army, plaguing it from the van, from the flanks, and from the rear, moving his militia units ahead of Percy, constantly replacing units as Percy passed them with new units as they arrived on the scene, then quickly bringing up the other units. He would put Percy's army inside a murderous, nonstop, full-scale attack all the way back to Boston. Percy's men weren't going to have a moment's respite for the rest of the day.

General Heath detailed his combat plan to the commanding officers of the various militia units and earned their eager acquiescence, for this was the type of fighting the veterans of the French and Indian War were familiar with. Under Heath's direction, militia units were quickly redeployed according to his plan, then resupplied with ammunition, watered, and fed. They then waited for Percy to resume his march. On the fifteen-mile stretch of road from Lexington to Boston, the duel between Heath and Percy was about to make military history.

For his part, Percy was an exceptionally gifted military officer, much more formidable than Colonel Francis Smith. And, allowing for the wounded and dead, his combined force was now twice as large as Smith's—more than 1,600 men—plus two very dangerous cannon.

At 3:15 Percy set off for home. He'd rearranged his force in three segments, with heavy flanker protection on both sides of the road pushing the militia back beyond musket range and with his two cannon protecting his rear, where he expected the heaviest action. He had the considerable advantage of being able to move his units within the line of march more quickly than the militia, which would have greater distances to cover to match his moves.

The ring of fire began to take its toll from the first. Despite the effective work of the flankers, regulars began to fall all along the line of march. Out on the flanks, General Heath kept up a steady flow of commands, moving his militia units with brilliant precision, keeping them from bunching up, and rapidly moving new units up to the front. Dr. Warren seemed to be everywhere, exhorting the men to still greater effort. But there was a price for the Americans. Despite the cover and the discipline, the militia was now also taking casualties all along the line.

Percy was deeply impressed with the leadership and discipline of the American militia. He surmised that he was up against a general every bit his match. From that moment on, like Colonel Smith before him, he could no longer think of the American farmers as a mob, a bunch of country bumpkins.

Lieutenant Frederick MacKenzie of the Royal Welsh Fusiliers concurred. His unit had the duty of covering the rear of the march for seven miles, often walking and firing backward to defend against the musket fire. His unit was relieved only when it began to run out of ammunition.

The wounded kept up by hanging on to horses and riding on the two cannon, from which they were unceremoniously dumped every time Percy put the two weapons into action.

Percy's troops, in a towering fury about this cowardly attack from behind stone fences, became desperate and filled with a murderous hatred. They had all now learned of the hatcheting of one of their own on Concord's North Bridge and believed they were fighting wild savages.

By the time Percy reached the town of Menotomy, there were more houses, less countryside, and the house-to-house fighting intensified. Here the regulars could actually get at the elusive provincials, and they paid them back accordingly. They entered houses along the roadway to route snipers and often killed everyone they found. They even killed unarmed civilians. They beat and battered, stabbed and shot, anyone they could get their hands on.

"All that were found in the houses," Lieutenant Barker later said, "were put to death."

Lieutenant MacKenzie noted in his dairy: "Some houses were forced open in which no person could be discovered, but when the column had passed, numbers sallied out from some place in which they had lain concealed, fired at the rear guard, and augmented the number which followed us."

He observed that the regulars were in a fury over the militia's tactics: "Our men had very few opportunities of getting good shots at the Rebels, as they hardly ever fired but under cover of a stone wall, from behind a tree, or out of a house; and the moment they had fired they lay down out of sight until they had loaded again or the column had passed."

In an oft-recounted episode, a seventy-eight-year-old crippled war veteran named Samuel Whittemore, armed with his musket, two pistols, and a saber, waited behind a stone wall to take his shots at the hated redcoats. When he got off five fast shots, the regulars charged. His musket dropped one, his pistols dropped two more, and while he was raising his saber, part of his face was shot away. Then he was bayoneted more than a dozen times and was left for dead. Even the doctor who attempted to care for him despaired. Yet he survived and lived another eighteen years, dying finally at the age of ninety-six, leaving behind numerous descendants.

In Menotomy, the ferocity of the American attack is told in the numbers: In that town alone forty redcoats died and eighty were wounded. Percy himself recounted that the provincials were so eager to shoot a redcoat that they moved to within ten yards, "though they were morally certain of being put to death themselves in an instant." The Menotomy home of Jason Russell stood right on the roadway. When the fleeing British came abreast, the fifty-eight-year-old Russell and a number of militiamen sprayed a deadly volume of fire on the troops. British flankers went around the back of the house and killed Russell and eleven others.

By this time, Percy, like Colonel Smith before him, had begun to lose control of his troops. They turned to plunder.

"The plundering was shameful," Lieutenant Barker said later. "Many hardly thought of anything else."

"Many houses were plundered by the soldiers," Lieutenant MacKenzie's account agreed, "notwithstanding the efforts of the officers to prevent it. I have no doubt this inflamed the Rebels and made many of them follow us farther than they would otherwise have done. By all accounts some soldiers who stayed too long in the houses were killed in the very act of plundering by those who lay concealed in them."

For Percy and his men, the worst still lay ahead in Cambridge. Several new regiments of militia appeared, eager to tear into that redcoated mass. Percy's fighting force was steadily decreasing, ammunition was dwindling, and there was no sign of a letup from Heath's moving ring of fire. It was now after five in the evening, and still they had miles to go.

Then Percy seemed to collide with a disaster. He was heading for the Cambridge Bridge, over the Charles River. To block his way, the Americans took up the boards. British engineers found the boards neatly stacked nearby and put them back down. The Americans came back, pried up the boards once more, but this time chucked them into the river. And there went Percy's planned escape route. He was about to be pinned against a riverbank.

Abruptly, unexpectedly, he turned his troops onto a track that led to the road to Charlestown; sheer desperation had broken him out of the ring of fire. The last of his cannon balls cleared the road ahead of him, and he had almost reached safety. But there was one more obstacle to be faced. Colonel Timothy Pickering had arrived with a large contingent from Salem and Marblehead—fresh, fully armed, and eager to finish Percy off.

But here Percy got a piece of unbelievable luck. Pickering, in a move that has been disputed ever since, held back his men. They were barred from firing a single shot as Percy passed them by and reached the safety of Charlestown Neck.

Brigadier General William Heath had been the right leader at the right place with the right tactics. Given the quality of soldier he had—battle-green, inexperienced farmers and mechanics, many armed with squirrel guns—his ring of fire had taken a heavy toll on the retreating, exhausted, outraged British regulars. He had turned in a military masterpiece.

Fearful as the casualties were for the regulars, had it not been for Percy they could have—should have—been much worse. As many as 3,500 militiamen had scored less than 300 hits because Percy's flanking parties and his two cannon kept driving them out of range. Only the courageous defense by Lord Percy had prevented complete annihilation. His tactics had been brilliant.

If the militiamen had managed to break open the column or if they had mounted roadblocks, they might have destroyed the entire British contingent. And had the late-arriving Pickering blocked the roadway and attacked, the British might not have had enough troops left for the Battle of Bunker Hill.

But while exultant over Heath's brilliant campaign and the superb fighting of the militia, many Americans were furious with Pickering. He had denied them the coup de grâce, a complete victory. They accused him of being a lukewarm Whig, hoping to avoid a final break with England. He claimed Heath had ordered him to stop. Heath, to the day he died, emphatically denied having given that order.

For the farmer-fighters, the battle had been an act of great courage. With their hardscrabble lives—the constant nearness of sickness and death especially among their children, days filled with plowing and the vagaries of wind, weather, and drought as they tried to produce enough food to last another winter, the risk of starvation if they didn't get the seed into the ground in quick order—these men had turned out to fight the largest, strongest army in the world. As a reminder, the smell of gunpowder would linger for days over the road to Concord and the strung-out litter of the battle.

For the British regulars, after nearly four hours of hell stretching from Lexington to Cambridge, the relentless pursuit and the remorseless slaughter had ended. The last of the rear guard of Royal Marines entered Charlestown led by Major Pitcairn. He had been at the head of the contingent to Concord and now he was the final one to reach safety. "Lord Percy looked at his watch and noted that the hour was past seven o'clock."

One of the finest military units in the world—some 1,800 men—had been humbled, decimated in a shooting gallery slaughter by thousands of New England farmers. Some 73 soldiers were dead. Another 174 were wounded and 26 were missing (captured, in most cases). The casualty rate among British regulars was nearly 10 percent, while the casualty rate among the 3,500 militia was much lower—less than 2 percent. For the families and friends of the militia casualties there was also much to mourn over—49 dead, 39 wounded, 4 missing.

Under the protective guns of the HMS *Somerset,* at anchor in the Charles, General Gage's badly battered troops collapsed with exhaustion after the worst night and day in their lives. One soldier noted: "I never broke my fast for forty-eight hours, for we carried no provisions. I had my hat shot off my head three times. Two balls went through my coat, and carried away my bayonet from my side." It was deep dusk, and shortly it would begin to rain on the place where the regulars had come to rest: Bunker's Hill.

Back in Boston, just as his partner, Benjamin Edes, was restarting his newspaper in Watertown, John Gill and Edes's son, Peter, were following the crowds flowing out of Boston as the first rumors of the fighting circulated in the city. A band of Admiral Graves's seamen seized them and marched them to prison. Booked there on a "trumped-up charge of possessing firearms," they were thrown into Boston Gaol.

To obtain their release, both men were required to post bail and agree not to leave Boston. But by June 19 Peter Edes was back in the jail and spent nearly four of the most harrowing months of his life under the unspeakable conditions of an eighteenth-century British army prison, which the historian Bruce Lancaster characterized as "indescribable": ". . . a quick death on the battlefield was often preferable to the interminable agony and brutality common in the prisons of both sides." Starvation, wasting diseases, savage beatings, the complete absence of even the most primitive medical attention—such conditions prevented all but a few prisoners from emerging alive.

Thus Lexington had broken up a famous propaganda partnership. Edes and Gill had spent years of Sunday nights—sometimes whole weekends, and often overnight into Mondays—in their grimy newspaper office, dispensing a steady invective that eventually set the whole colony in emotional flames.

Often surrounding the press with Edes and Gill were Samuel

Adams, James Otis, Dr. Joseph Warren, Josiah Quincy Jr., Benjamin Church, and their cohorts, all writing and editing articles, drafting jeremiads against the latest assaults by Crown and Parliament.

John Adams remembered spending one entire Sunday evening watching while Otis, Samuel Adams, William Davis, and Gill prepared the next day's newspaper: "—a curious employment, cooking up paragraphs, articles, occurrences, &c., working the political engine!"

But Edes was not finished. As the troops were sprawled on the ground on Bunker Hill waiting for water transport back to their barracks, Edes was pegging type for his next edition of the relentless *Gazette.* He would have a lot to say.

Riding into Worcester at seven P.M., Isaiah Thomas, publisher of the *Massachusetts Spy,* had come to a monumental decision. He would reassemble his press and, as his first order of business, write and publish his emotion-packed version of the battles of Lexington and Concord. Because of the battle, he now wanted his readers to understand that the time had come for a new understanding with England. He was about to present an idea that would shock his countrymen in all the colonies, even the most radical.

In Boston, after waiting all night and all day, fifty-four-year-old Lieutenant General Thomas Gage first saw his returning troops at a distance across the Charles River that evening as they struggled up Bunker Hill, disheveled, faces powder blackened, exhausted, burdened with their wounded. The general, as well as the shocked spectators who crowded the top of Boston's Beacon Hill, could also see the last of the military action, "the muzzle-flashes twinkling like fireflies in the gathering darkness." While watching the boats rowing over to Charlestown with fresh replacement troops and returning with loads of wounded, General Gage had to be wondering how he would explain this to his king.

The entire countryside was now in turmoil. Unreasoned terror drove ordinary people out on the roads, fleeing with a few possessions on their backs or in carts, going anywhere anyhow in any direction, some to a pesthouse, others to the middle of a field, others flocking around their parsons—and some of those parsons themselves losing their nerve and fleeing for their lives. Everyone was trying to get out of the way of the fighting.

Gage's attack had caught an entire society at a crucial moment of its agricultural life, the spring planting, and ripped it open from stem to stern, exposing everyone in it to desperation, ruin, loss of everything they had worked their entire lives for.

If General Gage could have been granted two wishes, he probably would have wished first for a successor to take over his detestable command and second for the city of Boston, and its arch-fiend, Sam Adams, to be flung into the nethermost bowels of hell.

Of all American cities, the English hated Boston the most. Boston was the source of more tribulations, insults, confrontations, challenges, tarring and feathering, rioting, violence, murders, and obstreperousness than all the other colonies combined. According to London newspapers, Boston was a "nest of rebels and hypocrites." It was "obstinate, undutiful, and ungovernable," and it "had committed more atrocious Acts of Outrage than any other Part of the Colonies." It was "a canker worm in the heart of the colonies" and "a rotten limb which (if suffered to remain) will inevitably destroy the whole body of that extensive country." Britain traced all its troubles with its colonies to Boston.

One Englishman summed up his country's hatred for that city for all time: "I would rather all the Hamilcars and all the Hannibals that Boston ever bred, all the Hancocks and all the Sad-Cocks and sad dogs of Massachusetts Bay; all the heroes of tar and feathers, and the champions, maimers of unpatriotic horses, mares, and mules, were led up to the Altar, or to Liberty Tree, there to be exalted and rewarded according to their merit or demerit, than that Britain should disgrace herself by receding from her just authority."

At the same time, when Thomas Gage closed the port of Boston—on June 1, 1774, some ten months before—Boston had been one of the economic prizes of the British empire.

Almost entirely surrounded by water, Boston was actually three thriving ports in one. It was a busy fishing port swallowing up tons of cod and other fish, a whaling port, and a commercial seaport offering everything an oceangoing shipping industry needed—abundant wharves crowded with ships, ropewalks, sail lofts, ship's chandlers, warehouses, shipbuilding yards, and dozens of other services, including a large supply of longshoremen.

More than sixty distilleries were producing a river of rum, much of it sent to England. An abundance of breweries matched the distilleries with a river of ale. Boston was crowded with craftsmen, working men, carpenters, cabinetmakers, painters, bricklayers, farriers, blacksmiths, millers, ministers, businessmen, lawyers, legislators, crown officers—all thronging the narrow crowded streets lined with shops and open markets, farmers selling their produce, butchers, candlemakers, coffeehouses, taverns such as the Green Dragon and the Bunch of Grapes, which were always crowded in an age when drink—rum, ale, cider, wine—was consumed in suicidal quantities. Also waiting to serve Gage's troops was a troop of prostitutes, an army of patriotic Mistress Quicklys always ready for a quick trip above-stairs, eager to embrace soldiers and seamen and local citizens alike.

Boston was a city noisy with the sound of money: the rumble of barrels on gangways, the clatter of wagon wheels and carts, the *clip-clop* of horses' hooves, the caulker's hammer, the ringing anvils of blacksmiths and farriers, the squeals of pulleys, the cries of lobstermen pushing their ladened barrows and of oystermen with bags over their shoulders.

And everywhere in the air were the smells that proclaimed furious human activity—the odors of pitch, of horse droppings, of the flung contents of chamber pots, of rotting garbage, of rotting fish, of dead animals, of exotic spices, of burning charcoal, of newly made rope, of rum distilleries and ale breweries, of cooking, of smokehouses, of tea and coffee and spices.

Moralists such as Sam Adams, dismayed by the endless abundance of luxuries that flowed into the city from England—fine quality plate, expensive clothing, horse-drawn carriages, china, furniture—contended that the moral fiber of the colony was being destroyed. Nevertheless, each year the people of the port of Boston processed goods by the millions of pounds, including lumber, ships' spars and masts, fish, rum, and slaves. Seventeen thousand Bostonians had created a moneymaking beehive.

Then General Gage shut it down, and the citizens reacted as though he had killed a living being. As the last cargo ships fled the port, Boston shops were shuttered, people appeared on the streets in mourning clothes, and church bells all over the city solemnly, slowly tolled. Boston was holding its own funeral.

The other colonies were equally funereal—colonists filled their churches, intoning prayers for Boston. In Philadelphia, the clappers of the bells of the Church of England were mournfully muffled, while in Williamsburg, the House of Burgess declared a day of fasting and prayer—despite the furious objections of Governor Dunmore who retaliated by promptly closing the House. After ten months, Gage presided over a corpse, a once-great seaport rotting on its empty piers.

For those ten months no shipping had moved in or out of Boston. The king, demanding payment for the tea cast into the harbor by a handful of "Mohawks," was punishing all of New England. Every kind of business had suffered, and many went bankrupt, never to reopen, ruining Loyalists and Whigs alike—a lifetime of labor lost. Hordes of idle seamen and longshoremen stood around the empty docks drinking and getting into trouble. Off-duty British soldiers were attacked on the streets.

The port's closure was even ruining businesses in England that traded with the colonies, affecting the empire's economic health.

The other colonies had sent food to sustain Boston's starving inhabitants. And although thousands of men were desperate for work, General Gage couldn't get a single Massachusetts laborer to drive a single nail for the barracks he badly needed to build for his ill-housed troops. To get the barracks built, he had to import Loyalist laborers from New York and have them carried to Boston on British warships.

But still Sam Adams blocked every move to pay for the tea. And so the situation had festered, month after month, until General Gage had made his disastrous move on Concord.

The most significant—and unintended—consequence of closing the port was the last thing George III wanted: He had driven the bickering, divisive, parochial, suspicious colonies into one another's arms. The first steps toward unified action had been taken in September 1774, four months after the closing of the port, at the First Continental Congress, to which every one of the thirteen colonies except Georgia had sent delegates. Leaders of the twelve colonies—who had never met before, who traditionally distrusted each other, and whose respective colonies were as separate as thirteen nations—sat together for the first time in Philadelphia in meeting halls and in taverns and in coffeehouses, looked each other over, and began to explore the possibilities of commonality.

Did a fat, sententious lawyer from Braintree, Massachusetts, named John Adams have anything in common with a wealthy, slave-holding plantation owner from Virginia named George Washington? Yes, he did; and much more than the two men might have thought, especially on the subject of their royal master and his grasping Parliament.

Worse, in the king's view, was that from that Congress had come more unified bellows of defiance than ever before. They politely petitioned the Crown in the most respectful terms, yes, but they also passed resolutions stating clearly that Parliament did not have the power to tax the colonies, that the colonies were not going to submit to being taxed, and that this stance was not negotiable.

General Gage was now pressed with a number of monumental problems. His title as royal governor of Massachusetts was a hollow scepter. His authority extended no more than a few inches beyond the nearest British bayonet—in all, a few streets in Boston. Sam Adams and his network of radicals throughout Massachusetts had effectively bottled up the British army in Boston and now had driven in the cork; Gage was trapped in the city he had been sent to discipline. With reinforcements from England weeks away, and a decimated and sick garrison, he didn't know what to do next.

How would he find food to feed his troops and the large and growing number of Loyalists who were streaming into the city, eager for the army's protection against the marauding rebels? Many had lost everything, having only the clothes on their backs. They would be completely dependent on Gage. Some 6,500 Tories had fled the radicals' tar and feathers, while 14,000 Whigs left Boston.

Then there was disease. A fever called "throat distemper" (possibly diphtheria) was raging through his garrison and had already killed several hundred, including Gage's confidential secretary and brother-in-law, Samuel Kemble. Dysentery, spreading rapidly, was also claiming a growing number of victims.

In addition, there was the frightening threat of "malignant spotted fever" (possibly typhus). Troops from a royal Irish unit were still quarantined aboard a ship in the harbor because they had arrived from Ireland with it. If the fever escaped from that ship, the overcrowded city would be a prime breeding ground for it.

Even the city's drinking water was foul—and might have contributed to the spreading diseases.

The city was the worst possible place to mount a fight—either offensively or defensively. Tens of thousands of armed colonists were coming from all over New England, crowding every road that led to the city, just as they had the previous September. With any sort of leadership, they could attempt to take the city, particularly if they could find a battery of cannons. Hell, it is written, is the truth perceived too late.

Compounding his store of troubles, Gage was faced with the most dismaying personal problem of his life. He had been betrayed. Someone in the inner circle of his military administration had warned the rebels of his planned raid on Concord. And he was sure he knew who that person was. Because of that betrayal alone, his military career faced extinction.

Regardless of personal feelings, Gage had to act. He had to answer three pressing questions. What to do about the traitor? What should his army do now? And what were the Americans going to do now?

General Thomas Gage needed a plan—and he didn't have one.

In Boston, the Honorable Thomas Flucker, Royal Secretary of the Province of Massachusetts, one of the wealthiest men in America, owner of a vast piece of the future state of Maine, and at the time an outcast in his own colony, had ample reason to be terrified by the sudden turn of events.

From a vantage point on Boston's Beacon Hill, Flucker, with a sinking heart, watched the battered troops come home. He, like the rest of the beleaguered Loyalists inside the city, was quite aware that the rebels could swarm in and gleefully annihilate the Loyalists unless they had an effective army to protect them. And having already lived through a nightmare of terrorism and torture for remaining loyal to the king, they had ample justification for their fears.

A Massachusetts-born American patrician and a determined Loyalist, Flucker himself had seen many of his associates driven from office and chased by Boston mobs wielding clubs and axes. When he wasn't looking at the troops on Bunker's Hill he could look with equal dismay in the other direction, down the long causeway of Boston Neck, to see many fellow Loyalists in full flight, streaming urgently into the city from the mainland, bringing only whatever they could carry or cart, having hastily abandoned their homes and a lifetime of possessions, friends, and relatives. Seeking the shaky protection of Gage's British army, they were becoming exiles in their own country.

Over the years, and most particularly in the previous year, the mob, led by Sam Adams with his messianic vision of a cleansed,

Puritanical, Bible-based city, had attacked hundreds of Loyalists—royal governors, ministers, prominent businessmen, judges, lawyers, legislators, customs officers, the surveyor of the port of Boston, and the register of admiralty, as well as farmers and mechanics, a store of talented, gifted, enterprising, educated, civic-minded people that the still rawboned colony could ill afford to lose. Adams's gangs had tarred and feathered many, burned their homes, and then chased them—and their wives and children—into the pale of the British army in Boston. Many of these American Loyalists would scatter and never return, would never see family and friends again, would live in the half-world of exiles among foreigners until the day they died, talking with their last breaths of "home."

And now, with the British port closed, Boston's once-thriving population of 17,000 would dwindle within three months to a starving, fever-racked 7,000, many of them homeless Loyalists.

Secretary Flucker, looking at the destroyed Loyalist families around him that day, could wonder how George III and his cabinet expected the British army, with a few cannon and muskets, to control not only all of Massachusetts but all of British America. The same British army had not been able to protect these—the most loyal subjects of the Crown—from attacks by mobs on just a few streets of Boston.

Flucker was particularly dismayed on this day because of his daughter. The previous June, Flucker had found himself the reluctant father-in-law of the notorious Henry Knox—an ardent Whig, the friend of Sam Adams, John Hancock and Paul Revere, the proprietor of Boston's London Bookshop, and a despiser of Loyalists. Flucker had tried to stop his headstrong second daughter, Lucy, from marrying Knox, and in an ironic prediction he warned her that if she married the young bookseller she "would be eating the bread of poverty and dependence." Flucker would become even more dismayed when his son-in-law, as Washington's artillery general, made a vital contribution to the defeat of Britain. But at the moment, his daughter was packing to leave with Knox, who intended to join the rebel government in Watertown. Saddened by the parting, Flucker had to wonder when and whether he would ever see his daughter again.

Thomas Brattle, a wealthy Tory, worried about the ability of the army to fight off the rebels' growing numbers. He had barely escaped with his life from an angry mob of 4,000 who, believing that he had participated in General Gage's seizure of 250 barrels of gunpower in September 1774, had chased him from his Cambridge mansion on foot for miles to the safety of Castle William, a fortress island off Boston. Now he could see armed men from all over New England standing on the shores across the river and looking hungrily at Boston. He had to doubt that General Gage and his army would ever restore him to his mansion—if it was still standing.

Inside the besieged city, many Loyalists were reminded of the depredations they had suffered at the hands of the mobs and how they had been driven to Boston.

Royal Chief Justice Peter Oliver, another victim of a waterfront mob, had watched the entire Massachusetts court system collapse. With gangs roving up and down the streets and in the courthouses for months on end, the forty-one-year-old judge had been unable to impanel a jury that would convict, had been often unable to impanel a jury of any kind, had found that in some juries the "mob leader was chosen to sit upon the grand jury appointed to investigate the riot in which he had played a distinguished part," had been unable to find witnesses, had been intimidated and threatened, and had finally been driven from his courthouse as rebels set up their own illegal court system there. The only safety for him and his family was penniless exile.

His brother, Thomas Oliver, the lieutenant governor of Massachusetts, had been confronted by a crowd of 4,000 who surrounded his house and pressed their noses on the windows while a committee of five men demanded that he sign a resignation on the spot. He wrote across the bottom of it, "My house being surrounded by four thousand people, in compliance with their commands, I sign my name." The next day, he moved to the safety of Castle William and resumed his duties, but with little effect. No one would obey his rulings.

Daniel Leonard, a Tory lawyer, observed the battered army that returned from Lexington and asked himself if it was time to leave his native country.

For ten months the thirty-four-year-old barrister had been living from day to day in Boston with his family of eight. Leonard, a leading attorney, had originally made some of the most ardent speeches in the Massachusetts Assembly against Britain. But during the previous June (1774), finally sickened by the violence and excesses of Sam Adams's mob rule, he publicly sided with Governor Hutchinson and became himself a victim of that same mob.

A marauding throng of 2,000—usually summoned after a night of free drinking by the signal of long, shrill whistles along the waterfront—was sent by the radicals to Leonard's home in Taunton. When they discovered that he was not there, they stood on the green nearby, roaring curses and threats.

The next night the mob returned.

They saw light in the window of a second-floor bedroom, where Leonard's wife was in labor. Someone in the crowd fired a shot through the window. When the newborn infant was deemed at birth to be an idiot, the Leonard family blamed the shock caused by the threatening mob. Leonard quickly rounded up his family and, abandoning his mansion and all its contents, departed for Boston. Nearly a year later he was still there, not daring to return to his home and now thinking about moving to England. He did not feel there was a future life for himself and his family in Massachusetts among the uncontrolled mobs.

The lawyer's departure would rob Boston of one of its more colorful figures. From an old Boston family and wealthy through marriage, Leonard had been quite famous throughout the province for his foppish dress. "He wore broad gold lace round the rim of his hat," John Adams told his diary. "He made his cloak glitter with laces still broader, he had set up his chariot and pair and constantly traveled in it from Taunton to Boston. . . . Not another lawyer in the province, attorney or barrister, of whatever age, reputation, rank, or station, presumed to ride in a coach or a chariot."

Although Leonard was a good friend of John Adams, his ostentatious public style must have caught the disapproving eye of Puritanical Sam Adams on many an occasion.

But behind the foppery was a brilliant mind that demolished many of the radicals' favorite arguments. Leonard was long remembered for his participation in "the greatest print war in the Province's history." It was ignited when, under the pseudonym of Massachusettensis, Daniel Leonard, after he moved into the Loyalist colony in Boston, began writing a series of seventeen essays that appeared in the Loyalist newspaper, the *Massachusetts Gazette and Post Boy.* These essays were so brilliantly written and so trenchant in their attacks on the radicals' philosophic posture that they are read by legal scholars and historians to this day.

John Adams, unaware that the essays were being written by his old friend and now political enemy, praised them highly, "shining" as they were in the midst of lesser Tory writing of the time, "like the moon among the lesser stars." They were, Adams said, "well written, abounded with wit, discovered good information, and were conducted with a subtlety of art and address wonderfully calculated to keep up the spirits of their party, to depress ours, to spread intimidation, and to make proselytes among those whose principles and judgment give way to their fears,—and these compose at least one-third of mankind."

Under the pen name of Novanglus, Adams eagerly wrote responding essays, possibly among the most inspired of his voluminous writings. The war of words was eagerly read and debated in coffee-houses and taverns right up until several days before the battle at Lexington and Concord, when the last Leonard essay appeared.

John Adams was not to discover for forty years the real identity of the author of the essays he so highly admired and attacked.

In Boston, Judge Samuel Curwen of the Admiralty Court in Boston, one of Sam Adams's most hated targets, was struggling with his decision to leave the country rather than put up with any future depredations of the Boston mob. While fearing for his life, he was also dealing with a ferocious domestic problem. His wife was opposed—violently opposed—to leaving. Her objection, he knew, was based on her pathological fear of the ocean. Malicious gossip said that was the true reason for his wanting to go to England—to leave her behind.

Edward Stow of Boston, "for being loyal to the King," became a victim of a mob that slathered his house with excrement and feathers, then, in a murderous attack, broke his head "with a cord wood stick."

A elderly dockworker in Boston named Malcolm was trapped by a gang celebrating the anniversary of the Boston Tea Party. Accused of being a Tory, he had one arm dislocated as the crowd tore off his clothing, stripping him "stark naked" in the bitter-cold night. They covered him all over with hot tar, then feathers. They trundled him through the icy streets in a cart, to the roaring delight of thousands of laughing spectators. His tormentors, using clubs, beat him out of the cart, then battered and pummeled him back in again. When he refused to curse his king, they put a noose around his neck and threatened to hang him. They probably would have been too late: "Several doctors said that it was impossible for the man to live because the flesh had literally been torn off his back."

In Boston, the Reverend John Wiswall and his two sons were trying to deal with their still fresh personal grief when they saw the returning army. An Anglican minister, Wiswall had watched the rising disorderliness in his parish over a period of months with astonishment. It was, he said, "a discontent bordering on madness . . . a leaven that spread from one town to another in Massachusetts." His fellow countrymen were "too free and happy . . . to be content with . . . happiness."

Reverend Wiswall himself became a victim of that madness just a few weeks before the Lexington and Concord battle when his friendship with British naval officers was noted. Wiswall found himself exposed to "the malice and rage of a lawless rabble" that so terrorized him and his family at gunpoint they fled to the safety of Castle William, where his wife and daughter promptly became infected with a violent intestinal disorder that was raging through that greatly overcrowded fortress. Both died. The bereaved Wiswall and his two sons would sail for England in January 1776.

Also in Boston, Timothy Ruggles, former Speaker of the Massachusetts Assembly who had presided over the Stamp Act Congress in New York City, had an idea. Why not use the many Loyalist young men from among Boston's Tory population of 6,500 as a military force? There were several thousand of them "in a hot angry mood," eager to fight. He decided to suggest this to the general. But the shilly-shally Gage snubbed him, evinced no interest in this cadre of ready young men, and thereby missed an opportunity to add several thousand able bodies to his slender ranks.

Ruggles had good reason for wanting to fight. Under Parliament's Intolerable Acts that closed the port of Boston, London had declared that the Massachusetts Council would no longer be elected by the House of Representatives. Instead, it was replaced by thirty-six Mandamus Councilors appointed by the crown. Accepting one of those posts was not the wisest choice Ruggles had ever made. He was visited by a gang of his fellow citizens who seized all the weapons in his home, poisoned a blooded British stallion worth more than the lifetime incomes of many of his visitors, and then put his son Timothy under house arrest. Like all the other thirty-six appointees, Ruggles hastily resigned and carried his family to Boston.

Also from Worcester, Timothy Paine, the wealthiest man in the town and another mandamus appointee, found his home surrounded by thousands of people who impelled him to resign, then forced him to pass "bareheaded through the crowd" as he read his resignation aloud. He fled to Boston about the same time as the Ruggles family.

Mob rule had already spread far beyond the Boston area and continued to inspire depravities by the rebels.

In Barnstable, Massachusetts, Abigail Freeman, a widow, was tarred and feathered for being a Tory.

In the middle of the night, with no warning, Samuel Jarvis, his wife, and their four children were stripped naked by a mob, then were ferried across Long Island Sound. Forced to wade ashore, they were then tarred and feathered.

In Rutland, Massachusetts, Daniel Murry was so terrified by 150 men carrying clubs that he not only resigned on the spot as mandamus councilor but also fled as fast as he could to the sanctuary of Castle William.

Not satisfied with tarring and feathering William Davies, a mob then savagely beat him with muskets.

Peter Guire was seized by a mob that branded "GR" on his forehead.

A mob poured hot pitch on another Loyalist and threw him into a hog pen, where they rolled him in hog dung, then forced more dung down his throat.

In Newport, Rhode Island, Martin Howard, the author of several essays that had infuriated the radicals, watched a mob hang an effigy of him from a gallows and then burn it. The next night, they returned and smashed their way into his home, where they spent hours trashing the house and its furnishings, destroying walls, ripping up the floors, then, in a joyful finale, burning the destroyed building to the ground. During a savage beating, Howard escaped and ran from the mob to the harbor and the safety of the HMS *Signet.* He never returned.

Daniel Dulany, a famous and aging legal scholar widely respected on both sides of the Atlantic, was profoundly opposed to the British treatment of its colonies and said so in a brilliant paper that became a key instrument in Pitt's attack in Parliament on British policy. A reasoned and committed patriot, he endorsed a break with England and was hailed by the Whigs.

But when he condemned the unbridled violence of the street gangs, a furious mob accused him of being a Tory, seized all his property, and made such terrifying threats that the elderly man went into seclusion and never emerged.

The most notorious "mobbish" attack of all, one that cost the colonial cause many friends in England and shocked the capitals of Europe as well as the other colonies, was directed at one of the highest ranking officials in the colonies, the royal lieutenant governor of Massachusetts, Thomas Hutchinson.

A native-born Bostonian whom Adams loathed with almost unbalanced passion, Hutchinson was in his home on a breathlessly hot night in August when a gang of 500 men was formed up on the waterfront. Summoned by the riot signals—the bell, the horn, the war whoop—they tumbled out of their favorite taverns and moved through the city, their raucous cries carrying into the closely shuttered houses of frightened residents.

After breaking into the home of the comptroller of customs, one of Adams's special targets, the mob got into his casks of wine and drained them. Now more drunk than sober, singing and bellowing, brandishing axes and clubs, the mob flowed toward the northern end of Boston to Garden Court and surged through the gateway to Governor Thomas Hutchinson's three-storied mansion.

Hutchinson, a widower, his children, and their aunt who was helping to raise them, had been at dinner when a neighbor urgently warned of the mob's approach. Hutchinson was determined to face the mob, but his young daughter threatened to stay with him if he didn't escape immediately. By slipping out his back door with his family just as the mob stormed up to the front door, Hutchinson undoubtedly saved his life.

Roaring for Hutchinson to appear in front of them, the wild mob trampled over the lawn and shrubbery, stumbled up the steps, smashed open the front door, then poured inside, whooping and yelling. They splintered doors and panels, shattered windows, and soon filled every room of the house up into the attic with furious chopping and hacking. They lugged a dozen wine barrels out of the cellar and soon were tossing back noggins of imported wine to help them with their drunken frenzy.

The mob battered and splintered wall paneling, chairs, tables, and picture frames, smashed glass and crystal, chopped furniture, ripped up floors. They carried the debris, including large pieces of furniture, over to the smashed windows and littered the grounds, yet there was much more debris on the floors inside, including smashed wine bottles, china, plaster, furniture, stuffing, glass, and shredded curtains and rugs. Plates were shattered; curtains and chair upholstery were slashed. Old family portraits of great historic value were cut to ribbons, children's dresses ripped to rags and dumped from the windows. They slit open mattresses and cushions, threw, kicked, and flung feathers through the rooms, then shook a blizzard of feathers out of the windows. Hutchinson's lifetime collection of priceless books and manuscripts, on shelving from floor to ceiling, was destroyed, chopped, ripped, trampled, and pitched from the windows. Even the governor's manuscript pages of *The History of Massachusetts,* years of scrupulous research and scholarship, went fluttering through the darkness to be scattered along the roadway.

Garden trees were chopped down, then chopped up. The flower beds were trampled by hundreds of feet. Men swarming on the roof pried the cupola loose and tumbled it to the ground. The roof slates were pried loose and scaled into the darkness.

Part of the mob went looking for Hutchinson himself and spent the night pounding on doors and searching homes in a rage. They failed to knock on the one door that would have led them to him. In an act of great courage, Hutchinson's young niece, Hannah Mather, had led him through back lanes in the darkness to the house of Thomas Edes, where he spent the rest of a sleepless night.

The pillaging, stealing, looting, and smashing went on without stop until dawn. With daylight, the town could see trails of dropped loot up and down the streets of Boston.

Hutchinson later itemized the damage: "Besides my plate and family pictures, household furniture of every kind, my own children's and servants' apparel they carried off about £800 sterling in money and emptied the house of everything whatsoever except a part of the kitchen furniture, not leaving a single book or paper in it, and having scattered or destroyed all the manuscripts and other papers I had been collecting for 30 years together, besides a great number of public papers in my custody."

The next morning, Hutchinson, who was also chief justice of Massachusetts, appeared for the first day of Superior Court. In the courtroom crowded with spectators, before four robed and wigged judges, Hutchinson appeared in just his shirtsleeves. Tearfully, he apologized for his condition. These were his only clothes. His family had just what they wore on their backs.

"I call God to witness that I never in New England or Old, in Great Britain or America, aided or supported what is commonly called the Stamp Act, but did all in my power to prevent it. This is not declared through timidity; I have nothing to fear. They can only take away my life. . . . I hope all will see how easily the people may be deluded, inflamed, carried away with madness against an innocent man. . . . I pray God give us better hearts!"

Adams's waterfront gang had gone too far this time. With the ruined building half buried in debris on full view in front of the whole town, the residents were appalled by the wanton destruction of one of Boston's finest mansions. The shame spread across all of Massachusetts. News of it reached all the other colonies and cemented in the minds of many a disgust with the brutality of Sam Adams. The next town meeting filled Faneuil Hall to overflowing. "Every face was gloomy," the normally bloody-minded *Boston Gazette* reported, "and we believe every Heart affected."

Eight ringleaders, particularly "Captain" Mackintosh, leader of Boston's violent South End gang, were arrested and charged with capital offenses. But another mob, probably Mackintosh's own gang, broke into the jail and released the eight.

They were not rearrested and were never brought to trial. Governor Hutchinson, driven from office and replaced by General Gage, later took his family to England and died in exile there.

Sam Adams had eliminated the leading Loyalist in Massachusetts and had his revenge on Hutchinson for humiliating his father, Samuel Adams Sr., many years before by maneuvering him out of the Governor's Council. If Adams had one regret, it was that he had not seen Hutchinson completely humbled by being tarred and feathered, a fate the ex-governor escaped only by slipping out of his back door and down a dark lane.

* * *

As the siege of Boston wore on and conditions worsened, John Andrews, a Boston businessman, wrote to a correspondent: "We have now and then a carcase offered for sale in the market which formerly we would not have picked up in the street. . . . it is said without scruple that those who leave the town forfeit all the effects they leave behind." Since his business and inventory and home were worth "between two and three thousand sterling," Andrews didn't dare leave the city. Then Andrews reported in his diary the ultimate indignity: "August 1st. This day was invited by two gentlemen to dine on *rats.*"

When the weather turned cold, wood would become like gold, and the British soldiers were to be long remembered for tearing out the carved paneling and superb pews and seats from the Old South Meeting House. Stripped to the walls, the historic old building was turned into a British army riding school. The troops were to get some grim satisfaction from chopping down the century-old elm tree on the common—which had been turned by Sam Adams into the Liberty Tree—to warm British army bottoms. Houses, fences, trees, anything and everything was carried off by the soldiers trying to get warm. Andrews's barn and wharf were torn down, chopped up, and stuffed into army fireplaces "by order of General Robinson."

From a life of privilege and wealth, Chief Justice Peter Oliver, like the rest of the cowering Loyalists, was reduced to unspeakable conditions in the city. Oliver wrote to former royal governor Hutchinson, now in London, that even the rats were starving: "We are in the rotations of salt beef and salt port one day, and the next, chewing upon salt pork and salt beef. The very rats are grown so familiar that they ask you to eat them, for they say that they have ate up the sills already, and they must now go upon the clapboards."

In Braintree, John Adams, Sam Adams's cousin, had fallen into shock at the news from Lexington and Concord. In a tangled emotional state, mixing elation at the prospect of wresting free from England and dread at the prospect of a war against the greatest army in the world, John Adams mounted his horse and rode to Cambridge, where he visited General Artemus Ward and studied the militia camps. "The New England Army," in high spirits and vowing to push the British army into Boston Harbor, was hardly an army at

John Adams

all—unbuttoned, untrained, undisciplined, without food or proper housing.

From there he solemnly rode out on the littered roadway to Lexington and then to Concord, turning in at farms and homes to get firsthand viewpoints on what had happened. He felt great dismay when he saw the bodies of the dead young soldiers still unburied and when he witnessed the families of dead farmers weeping.

He rode into the hamlet of Lincoln, where the men of the Hartwell family loaded the bodies of five British soldiers onto an oxcart and carried them to the cemetery, believing one of them, with his hair tied in a queue, to be an officer. The scene was repeated for Adams all up and down the roadway. On Lexington Green, under the direction of Reverend Jonas Clarke, the dead were gathered up and placed in pine boxes ''made of four large boards nailed up,'' as his daughter remembered.

The journey became a defining watershed moment in Adams's

life. "It convinced me," he wrote, "that the Die was cast, the Rubicon crossed." At that moment the Revolution had begun for him. No more was there a road back into the British Empire. But the crossing of the Rubicon was also profoundly dismaying. He was so sickened by the carnage he observed on the Concord road and so awed by its significance that he took to his bed with brain fever. Emotionally, he still had to come to terms with what he had already accepted intellectually.

During his entire life, Adams's emotions had often lagged behind his reasoning, and this had caused him to go through the strangest of spiritual metamorphoses. As a boy he was shy, unsure, even timorous, fleeing from confrontation. There wasn't even a slight hint of the powerful, pugnacious leader that finally appeared. And he had shown little of that intelligence that was to light the colonial world. In fact, he disliked study so much he asked permission from his father, a farmer, to drop the study of Latin. His father agreed. "Try ditching instead," he told the boy. Two days of that, though, sent him hurrying back to his Latin text.

He was still lacking in social poise when he entered Harvard at fifteen; was still unassertive when he went off with his degree into the world inauspiciously—as a threadbare, hesitant teacher of Latin in a church grammar school in Worcester.

Awkwardness and immaturity dogged him even at twenty. He felt inferior, struggled at tasks others found easy, and yearned for charm and popularity, leadership and fame. He was pompous. Stiff. Irritable. Given to youthful moralizing and scolding. At least he had finally found a guiding light: ambition. But his immaturity would continue to betray him. His friends tended to laugh at his efforts at posturing. Robert Treat Paine joked about Adams's interest in Greek. Samuel Quincy was amused when Adams told him he wanted to be a great man.

Seeking money and status, Adams apprenticed himself to a gifted Worcester lawyer, James Putnam, and spent two years studying law by night, teaching Latin by day. He often debated with his tutor, learned how to construct arguments, and although a decided conservative, imbibed some of Putnam's liberal ideas. But hardly all. He believed that the vote should not be given to the mentally incompetent, the insane, the criminal, people in debt, children, or women.

The eminent lawyer Jeremiah Gridley, after careful observation, gave the socially cross-footed youth two bits of advice: "One is to pursue the law itself, rather than the gain of it. . . . The next is not to marry early."

To his great frustration, as a lawyer he was at first regarded as barely competent. His income was modest, the work long. Everything was great labor. Nothing came easily. He lost his first case and felt that he was lagging far behind his peers. His self-assessment in his diaries was brutally honest. He berated himself for being sobersided, intellectually vain, too quick with his sharp tongue. He agonized over his view of himself in society; when he attempted light banter with others, he came off as not witty but silly. And he was tormented by his terrible self-consciousness around women. Excessively sensitive, inclined to be quarrelsome, he often brooded over slights. And his pride of opinion kept him from changing his mind. He did so only grudgingly—and often too slowly.

John Adams had a very prickly disposition that grew spikier as he aged. His harsh judgment of himself never mellowed, and his assessments of others were often vicious.

All these traits were handicaps in his career.

Hesitantly, he courted Abigail Smith, daughter of a Congregational minister. Not physically robust, she was well educated and an avid reader of poetry, fiction, and great literary works. She was an exceptional writer and corresponded extensively with her friends. Adams had found his intellectual match.

But Abigail Smith had reservations about him. Occasionally she felt uncomfortable in his presence. In truth, he was not much of a catch. Not from a distinguished family. No money. Unexceptional prospects. Halting. Unsure of himself. He would have to court her for four years before they were married.

Abigail's father was none too pleased with her choice, but Adams gained greatly from it. She was a superb confidante and adviser, and through her support Adams became more poised and self-confident.

Bit by bit, the great man appeared. Some of his pomposity was rubbed off by the earthiness of life around him. While defending a man in Salem against a charge of bastardy, Adams put his client on the stand. The accused helped win his own acquital when he asserted "I fucked once, but I minded my pull backs. I swear I did not get her with child."

The landmark Otis case against the Writs of Assistance politicized Adams. In his famous oration, Otis was the first to enunciate the thesis that laws that violated the British constitution were void and should not be obeyed. Young Adams took that to heart. He was becoming more sure of his philosophic view of the world.

In time, experience began to harden him; his legal career prospered, and his reputation spread. He first became famous throughout the colonies for his defense of John Hancock against charges of smuggling. Adams contended that the law containing the tariff of seven pounds per ton for Madeira wine was passed by Parliament in 1765 without Massachusetts's representation. It was, therefore, unconstitutional—"made without our consent." He told the court, "My client, Mr. Hancock, never voted it and he never voted for any man to make such a law for him." It was, therefore, a case of taxation without representation.

But this argument was hardly new to British ears. In fact, it wasn't even American in origin. The Irish had been demanding—unsuccessfully—the end of taxation without representation for thirty years. The plea had no more effect in Adams's hands. The reason the court eventually folded its case against Hancock was far more practical—over 350 witnesses to the unloading of the smuggled wine developed amnesia.

Yet there was a new boldness now in Adams, a new sureness in his angry tread. He felt he knew exactly where he was going.

From being merely politically interested, he became politically active when he joined in the attack on the Stamp Act. His four notable articles in the *Boston Gazette* presented the colonial case against the Stamp Act so effectively they were reprinted in London and widely read there. Each propounded the same theme: the Stamp Act was illegal because Massachusetts had not assented to it and therefore was not required to obey it.

Adams came to more widespread prominence when he drafted the instructions for the Braintree representatives to the Massachusetts Assembly to strongly oppose the Stamp Act. Many other towns copied his instructions for their representatives, and he soon found himself among the recognized leaders of the Massachusetts radicals.

As his leadership qualities became more pronounced, others were showing a greater willingness to follow him. Almost against his will he was being pulled ever more deeply into Boston politics. But he hated the mob and shunned taking a role in Boston's town meeting. Even though he reacted angrily to British policy and practices—especially when British troops landed in Boston—he wished to stay out of public life. His reservations grew when he observed how skillfully his cousin Samuel Adams could sway the popular mind. He wondered how Sam could place such great confidence in the common man when he could so easily manipulate and enflame him.

But the times conspired against John Adams's aloofness. His stature grew among the colonies during the infamous trial over the so-called Boston Massacre, which, the town whispered, had been manipulated from first to last by his cousin.

John Adams had taken on a ferociously unpopular case. Defending a British army officer and a group of soldiers against the charge of having murdered Boston citizens in the streets required great moral and physical courage. He won an acquittal for the officer and most of the soldiers. Only two of the soldiers were convicted, and their only punishment was the branding of their thumbs when they pleaded an old law of benefit of clergy. Adams's courage and integrity were recognized by the voters the following year when he was elected to the Massachusetts House of Representatives by a vote of 418 to 118.

As he became more sure of himself and his philosophic view of the world, his temperament became more volcanic. In town meetings he often bellowed with fury at his opponents, hammering home his points by pounding his fist on his palm. He had become formidable. Frightening. He could not even read a book calmly. His 3,000-volume library bristles with his marginalia. "Pitifull!" he scrawls. "Thou Louse, Flea, Tick, Wasp or whatever Vermin thou art!"

During the debates in the Continental Congress, even on brutally hot days, the windows of Independence Hall were often closed when he spoke to prevent his bellowing from carrying into the streets of Philadelphia. He didn't discuss a subject. He attacked it. Even his writings, which were voluminous, were full of rage. Sharply reasoned, carefully structured, they persuaded many a mind during the wavering crucial days before Lexington.

As a mature lawyer, Adams's ambition was too evident. James Otis once said in a rage to Adams: "You don't care for anything but to get enough money to carry you smoothly through this world." But this was no longer the full picture of him. Adams had committed himself to a defiant posture against Britain that could have earned him not a smooth way through the world but a cobbled ride to the hangman's noose. His commitment to the colonial cause, forged in the fire of controversy, became inalterable.

As a public figure, Adams had no illusions about his popularity outside Massachusetts. The reasons he gave to Jefferson that he—not Adams—should be the draftsman of the Declaration of Independence are quite revealing: "Reason first, you are a Virginian, and a Virginian ought to appear at the head of this business. Reason second, I am obnoxious, suspected, unpopular. You are very much otherwise. Reason third, you can write ten times better than I can."

But when he learned to control his impatience and to carefully observe his opponents, he became "a master of the workable compromise and meaningful debate." Leading the floor fight in favor of Jefferson's Declaration, and convinced it was essential for the survival of the colonies, he was later given major credit for pressing it through a thicket of waffling opposition led by the terrifying predictions of John Dickinson. Adams, said Jefferson, "came out with a power of thought and expression that moved us from our seats."

In the midst of a remarkable collection of brilliant intellects rarely seen in any age, John Adams exhibited time and again one of the finest minds of his age. And yet he paid a high price. He was not a born revolutionary. He loved order, the rule of law, hated violence. Having decided on independence years earlier than most, he still accepted the idea with the greatest reluctance. He wrote to Abigail, "I go mourning in my Heart all the Day long." Yet once he accepted that eventuality, his unflinching efforts to bring about political freedom from England and his intellectual contribution to the political organization of the new country were matched by few others.

Jefferson said, "A man more perfectly honest never issued from the hands of his Creator."

In the village of Acton, several miles northwest of Concord, the day after the battle, Hannah Davis had received the body of her husband, Isaac, captain of the local Minutemen, who had been among the first to be killed at the North Bridge in Concord. She had to make plans for the burial. He had left her with sick children, several with rashes that were likely to prove fatal. She had a farm to manage and little means to get the planting done. Without planting there would be no food. Without food, winter could bring starvation.

In Boston, Henry Knox, the large and portly twenty-seven-year-old Boston bookseller—who weighed over 300 pounds, it was said—was busy packing. He was among the last few radicals still in Boston.

Henry Knox

Aware that Admiral Graves had shore patrols of sailors arresting radicals and suspected traitors, he hoped to evade Graves's posses by slipping through the British lines under cover of darkness. He would take with him his bride of less than one year, Thomas Flucker's portly, strong-headed daughter Lucy, and within a few hours report for duty at the headquarters of General Artemus Ward in Cambridge.

Knox was a prominent figure in Boston. Huge, powerful, and in his youth a well-known strong boy and head of his own gang, he was now always superbly dressed, a compelling conversationalist, and an indefatigable correspondent. On his left hand he wore his signature scarf or silk banner, which concealed two missing fingers, lost in a hunting accident many years before.

Knox was one of ten sons abandoned by a shipmaster father when Knox was nine years old. He grew up as a bookseller's apprentice, voraciously reading every book he got his hands on, and acquired an impressive education and a gentleman's manner. At twenty-one he had opened the London Bookshop, an intellectual center in Boston. Its stock of British military books was a magnet for many British officers, as well as for a number of colonists who felt it was time to learn about fighting a revolution.

It was from these books that he himself became a self-taught expert on artillery tactics and ordnance. A member of the Boston Train, an association of artillery enthusiasts who fired their own brass cannon, his knowledge of artillery was so profound that his father-in-law, Flucker, had brought to him an offer of a commission from the British army. Knox had turned it down. Hardly a Loyalist, he was in the thick of the radical movement in Boston, contributing articles to the radical press and spending hours in the Green Dragon Tavern in the company of Paul Revere and Sam Adams, imbibing treason.

By fleeing, Henry Knox would abandon his home and his bookstore. By accompanying him, Lucy Flucker Knox would never see her parents again.

Meanwhile, Dr. Joseph Warren, who had gotten out of Boston on the night of the eighteenth to participate with wild abandon in the rout of the British army, was busy on General Artemus Ward's staff in Watertown, even before the general had gotten there, planning the next military moves against General Gage.

Sitting next to him at the meeting of the Committee of Safety was Dr. Benjamin Church, who ranked almost as high as Dr. Warren in patriot affairs. Church had acquired a reputation for courage under fire: He had shown Revere blood on his stocking, claiming it was from a dying militiaman he had led in battle. Dr. Church announced to his fellow committee members that he was going into Boston the next morning.

He was being foolhardy, they exclaimed. He would be arrested.

"They will hang you if they catch you in Boston," Dr. Warren warned him.

Dr. Church said he was going anyway.

The other members remained concerned. Who was he going to visit? His mistress?

Dr. Warren now turned to the record of the battle. He wanted overwhelming documentation of the British army's depredations. So he issued an order to the justices of the peace in towns along the road to Concord to take sworn affidavits from eyewitnesses, testifying to what they had seen the British army contingent doing. He expected to get at least 100 depositions.

And with them, he would make history.

Boston had grown strangely quiet; the newspaper war—with the fulminations, the screaming insults, and the character assassinations—had stopped.

The only paper still on its feet in that city was the *Massachusetts Gazette and News Letter,* staunchly Tory, and it would survive only until the following year, when the British would evacuate Boston for Halifax, taking with them the whole beleaguered colony of Tories. With its customer base gone, the newspaper would cease publication forever.

The customary habit in the London Coffee House and the taverns—the Bunch of Grapes, the Green Dragon, the King's Head—of reading the paper aloud and either applauding or howling with rage was for a time broken. Out on the farms and in the small villages, reading by the fire under the ticking clock one's favorite dose of slanderous misrepresentations was one of the highlights of the week. For the first time in years, the radicals in and near Boston were confronted with the choice of reading a Tory slant on the news or doing without.

If the American Revolution was primarily a war of ideas, then the newspaper battle must be regarded as not a peripheral activity but as the central war of the whole period. A glance at their numbers shows the influence of newspapers. In ten years—from 1765, when the major confrontations with England really began, to 1775, on the eve of the Revolution, the number of newspapers in the colonies had leaped from twenty-one to thirty-seven, almost all of them radical.

It was an age of unparalleled and unbridled invective. The typical newspaper screed of the day often ignored the opponent's argument, preferring to attack the man himself using barely disguised references to its adversary. The level of abandoned polemic, vilification, and untrammeled libel was extraordinary. Typical was the comment of an editor in Boston who attacked three opponents with rapture: "What a wretched Triumvirate! A poor shiftless erratic Knight from Abington, a dunghill-bred Journeyman Typographer and a stupid phrensical Mountebank." The columns shook volcanically with such vituperative phrases as "braying Bretheren," "disordered Imagination," "a few desperate Incendiaries," "Court Scribblers," "detestable trash," "perjured traitor," "usurper," and "a monster in government." One newspaper called James Otis a "filthy skunk."

Many threats appeared, barely veiled and mainly from the Whigs. One savagely attacked a "Court" writer by warning him that there were those who "may be tempted to usher you into public notice, by breaking your bones . . ." Another attacked Thomas Hutchinson in the most abandoned language, calling the governor and his "crew, the greatest of criminals—more worthy to be delivered into the hands of the executioner than private robbers, assassins or murderers."

In an age when a single unfounded epithet could destroy a man's reputation, there was nothing funny to shocked readers about such completely unfounded attacks as this from the *Newburyport Essex Journal* on General Gage:

In truth, it's judg'd by men of thinking
That GAGE will kill himself a drinking.
Nay, I'm informed by the inn keepers,
He'll bung with shoe-boys, chimney sweepers.

The press distorted the news and also suppressed it at will. Significant events such as the vicious assault on customs inspectors by Sam Adams's mob were not reported. On that occasion, several inspectors had been locked in a cabin on board Hancock's ship while a smuggled shipload of wine was offloaded. With a crowd of over 300 cheering, the inspectors were then pitched headlong into the harbor. At the same time, Adams's mob seized both the collector and the comptroller of customs and hauled them through the streets with taunts and shoves and punches while behind them the collector's son was dragged along by the hair of his head. Adams, whose reputation for violence was already gamy enough, obviously didn't want the rest of the colonies to know any more about his strong-arm methods.

The then governor, Sir Francis Bernard, not appreciating the tremendous influence that newspapers were acquiring, foolishly dismissed Adams and his cronies as "Tavern Politicians" and "News Paper Libellers." He failed to see the terrible damage they were doing by undermining his royal government. His own standing was greatly diminished when they described him and his staff as "those guileful betrayers of their country."

Samuel Adams, under his pen name of Populus—one of many—hooted at the governor: ". . . there is nothing so *fretting* and *vexatious,* nothing so justly TERRIBLE to tyrants, and their tools and abettors, as a FREE PRESS."

In the colonial newspaper wars, honesty, truth, and civility were all casualties.

Perhaps it is not too strong to say that the gravest mistake made by Crown authorities in the colonies was in not shutting down the newspapers. Such was their influence that without them Samuel Adams might have remained just an ineffectual Boston waterfront agitator. Without newspapers, there might have been no revolution.

In Watertown, on the morning of April 19, when he received the first reports of the fighting, Colonel Joseph Palmer of the Committee of Safety scrawled an urgent announcement of the events and summoned Israel Bissel. He sent Bissel down the Boston Post Road to deliver the news to a string of towns south of Boston; his farthest destinations were New York City and Philadelphia. A resident of East Windsor, Connecticut, Bissel was an express rider by profession who traveled the routes between Boston and New York regularly.

Each town he passed through was left fulminating.

In Shrewsbury, Bissel alerted fat, slow-moving Artemus Ward, brigadier general and second in command of the Massachusetts militia. Racked with gallstones, the forty-eight-year-old General Ward struggled from his bed to his feet, climbed into the saddle, and in spite of extreme pain rode to Charleston. A militia veteran of the battle of Ticonderoga and a hardworking, efficient officer, he was slow of speech but quick to act. Assessing the situation in Charleston, he, unlike General Gage, knew exactly what to do. Gallstones notwithstanding, he promptly took over command of the thousands of milling militia from General Heath. To prepare for full-scale war, he set the rebel soldiers to digging fortifications.

Racing with the most important message he had ever carried, and within thirty-six miles of home, Bissel escaped injury when his first horse dropped from exhaustion and died just short of the Worcester meetinghouse.

On a new horse supplied to him in Worcester, Bissel set off once more, almost due south into Connecticut, shouting the news in every town he passed through. In his wake, bands of young men shouldered their muskets and began walking to Boston.

* * *

Down in Pomfret, Connecticut, at three in the afternoon, Bissel found the powerfully built, five-foot-seven-inch Israel Putnam. "Old Put," clad in a leather jacket and apron, was supervising a group of workmen building a stone wall on his farm. A legendary Indian fighter in Rogers' Rangers during the French and Indian War, and later commander of a provincial regiment under Amherst during Pontiac's War, the fifty-seven-year-old Putnam was an enormously popular figure, if not a living legend.

Chairman of the Connecticut Committee of Correspondence, soon to be one of Washington's generals, Putnam had added greatly to his enormous stature in New England when, after Gage's closing of the port of Boston, he personally drove a herd of 130 of his sheep 100 miles to help feed the city. Now, without pausing to change from his leather work clothes, Putnam issued an order for the Connecticut militia to be called out, then mounted and hurried to Cambridge, covering the same 100 miles on the same horse in eighteen hours.

Hastening farther south into Connecticut, Bissel reached New London by seven o'clock in the evening of April 20, rousing every town and village he passed through, talking to the crowds that came out, getting fresh horses, snatching fitful naps and bites to eat before resuming.

He left New London at midnight and rode west all night along the coast of Long Island Sound. At one A.M. he was in Lyme; by four his voice woke the whole town of Saybrook. Guilford, just waking, heard the stunning news at ten A.M., Branford at noon, and New Haven at noon on April 21.

One of the most elated New Haven residents was the thirty-four-year-old Benedict Arnold, apothecary and successful importer. Sunk into a soured marriage to the former Peggy Mansfield (she believed he had contracted a venereal disease in the Caribbean and would no longer sleep with him), and with three sons to raise and an importing business that the Crown's inflexible mercantile policy was destroying, he donned his uniform and summoned the town militia. With the characteristic aggressiveness and unstoppable force that was to gain him both admiration and disgrace, he then browbeat the town council into opening the town's arsenal, from which he armed his troops and set off for Boston, hungry for glory.

While he was away he would become a widower.

* * *

By this time other express riders were carrying Bissel's message all over the rest of New England. When the news reached Newport, Rhode Island, the effect was galvanizing. The newly reelected Loyalist governor, Joseph Warren, refused to call up troops to go to Boston on the plea that Parliament would revoke the colony's royal charter. The colonial assembly summarily turned him out of office and began making plans for war.

The day after the battle at Lexington and Concord, militia generals were arriving at Watertown hourly. The first staff meeting included Generals Ward, Heath, and Whitcomb, plus numbers of lower ranking officers. The top command was befuddled. It had an army thrust on it before it had a chance to decide what to do with it. There was a serious shortage of equipment and materiel, including "bayonets, powder, horses, clothing, and tents and commodious barracks."

General Putnam arrived the next day. But by then many of the Minutemen of Massachusetts and Connecticut had gone home.

On April 20, General Gage, to the dismay of some of his officers, withdrew all forces from Charlestown by longboat back to Boston. Convinced that the British would now shell the town to burn it down, almost the entire population of Charlestown, some 3,000 residents, began to pack up and flee. By June only a few hundred would remain. The expectations of the majority proved correct. In preparation for the battle of Breed's Hill, on June 17, Gage would raze the town, leaving it in ashes.

In Coventry, Rhode Island, Nathanael Greene, a rather tall and handsome thirty-three-year-old Quaker with a pronounced asthmatic wheeze, acted quickly when he heard the news, taking the first steps to try to mobilize Rhode Island's militia.

From tough hard labor in the family ironworks from early childhood, Greene was a physicially powerful man—despite having damaged one of his knees and injuring an eye in the process. Believing book learning from other than the Bible to be too sinful and worldly for Quakers, his father had summarily ended Nathanael's education in early childhood by dismissing the boy's tutor. But Greene had learned enough to continue reading on his own, and over a period of years he became largely and effectively self-taught.

His reading really expanded when, on a trip to Boston, he discov-

Nathanael Greene

ered the London Bookshop. He and Henry Knox were soon deep in conversation. Greene, stepping into a wider world of books than he had ever seen before, began buying selectively, reading history, bookkeeping, science, and law. He mastered Euclid by keeping a copy on his forge while he worked. He soon worked out a barter arrangement with Henry Knox, paying for his books with toy anchors he made on his forge.

In an age when books were expensive and rare, Nathanael Greene built up a library of over 200 volumes.

Becoming head of the ironworks at the death of his father, Greene became increasingly convinced that revolution lay ahead. After the Boston Tea Party, with no military experience and despite a strong pacifist background, he decided to master the art of warfare. Like Knox—and William Heath and Washington himself—he patiently plundered the contents of books to do so—Plutarch and Julius Caesar, Sharp and Turenne among others.

When the Independent Company of Kentish Guards was chartered in 1774, Nathanael Greene knew more about warmaking than anyone else in Rhode Island. He offered his services as an officer. The guards rejected him because of his stiff knee. He signed up anyway, as a private, which led to his being disowned by his Quaker Meeting. By the time of the battle of Lexington, the Kentish Guards had raised him from private to brigadier general.

Rhode Island's militia was well supplied and equipped by that colony's merchants—Christian businessmen in Providence, Jewish businessmen in Newport—but it was delayed getting on the road until early May by Governor Warren's opposition. Greene himself was chosen to lead the three Rhode Island regiments—complete with tents, wagons, supplies, and even artillery—to Boston, where he reported to General John Thomas at Roxbury. The following month, Congress commissioned him brigadier general in the Continental Army.

During the war, while he was on military service, he had his ironworks converted into a cannon factory. His troops got used to seeing him halt a march, make a cup of tea, and, slipping a book out of his pocket, become totally lost in his reading. He was destined to become the most brilliant of all of Washington's generals.

Greene is best remembered today for his wry comment following the battle of Bunker Hill. "I wish," he said, "we could sell them another hill at the same price."

Two days after the Lexington and Concord battle, Hancock and Samuel Adams remained in hiding in the village of Woburn. Most reports indicated that General Gage's army was back in Boston and showing no indications of another strike, so they felt it was time to continue on to Philadelphia. During the day they moved to the village of Billerica, where they would remain two more days. Their immediate destination was Worcester, where they planned to meet up with the other delegates on their way to the Continental Congress.

Sam Adams had still not recovered his abandoned trunk of clothing.

On Saturday, April 22, after resting overnight in New Haven, Israel Bissel reached Fairfield, Connecticut, by 4 that afternoon, three days after Lexington. His next major city would be New York which, under a good horse, he could expect to reach late the following day,

Sunday, April 23. He was totally unprepared for the reaction he would cause there.

With the effects of the battle three days before still evident in the town, the Massachusetts Provincial Congress re-assembled in Concord and determined that an army of thirty thousand men was necessary for the defense of the country. It then voted to raise an army of thirteen thousand six hundred troops as Massachusetts's contribution to that army.

Three days later, the Rhode Island Assembly voted to raise an army of fifteen hundred men. And the day after that, on April 26, Connecticut voted to raise an army of six thousand men.

Everywhere around Boston, men were digging fortifications.

By the afternoon of Saturday, April 22, General Thomas Gage knew he could delay no longer. He sat down to write the most important document of his career—his report to his king on the battle of Concord and Lexington. It was a heavy hand that held his pen.

The general was still stunned by the deep significance of the events. Colonials had fired on British troops. A unit of the British army had been decimated and defeated. Out of his expeditionary force of 2,000 men on the nineteenth, 73 had been killed, 174 were wounded, and 26 were missing. In all, he had suffered a stunning 273 total casualty rate—more than 10 percent—in that operation. His effective fighting force was down to about 3,500 men and dwindling, with fever and sickness taking a daily toll. Waiting for him in a ring around Boston were more than 20,000 armed colonists, with more arriving by the hour.

Gage and his senior staff still regarded the Americans as a loose band of colonists spontaneously taking advantage of a military situation. It would be some time later when they finally understood that, right under their noses, the radicals had raised an entire army of Massachusetts militia in every city, town, and village, trained it, and placed it under the command of five generals—William Heath, Artemus Ward, Jedediah Preble, Seth Pomeroy, and John Thomas.

British army officers, not understanding the significance of what was plainly in front of them, watched with wry amusement as farmers and mechanics and businessmen struggled to learn musket drill and marching.

"It is a curious Masquerade Scene," wrote an English visitor to a friend, "to see grave sober Citizens, Barbers and Tailors, who never looked fierce before in their Lives, but at their Wives, Children, or Apprentices, strutting about in their Sunday Wigs in stiff Buckles with their Muskets on their Shoulders, struggling to put on a Martial Countenance. If ever you saw a Goose assume an Air of Consequence, you catch some faint Idea of the foolish, aukward, puffed-up state of our Tradesmen."

But they failed to sense a key point picked up by a colonial woman in Wilmington, North Carolina, as she, too, observed the sweating, awkward movements of 2,000 militia marching in review. While totally unimpressed with their military appearance, she noted ominously, "The worst figure there can shoot from behind a bush and kill even a General Wolfe."

General Gage and his army could condemn the Americans for not fighting by the rules. They could call them savages, especially considering the much-magnified account of the "scalping" on the North Bridge. They could minimize the casualties they suffered at the hands of the Americans. But they could not hide the most significant fact: A collection of farmers, artisans, businessmen, and professional people had organized into an army and with shockingly effective leadership and military discipline had made nearly 2,000 British regulars run for their lives through seventeen miles of Massachusetts countryside.

On the surface, General Thomas Gage, at fifty-four years of age, would appear to have been a perfect choice for the assignment his king had given him.

The general was married to Margaret Kemble, a wealthy American heiress from a very influential New Jersey family, by whom he had had six sons and five daughters. A career officer with more than thirty years of military experience, he had spent twenty of those years as an officer in America and knew most people in power in the various colonies—including Washington, with whom he had served at Fort Pitt under Braddock in 1755 and with whom he continued a social friendship. What British general could claim to know how to handle the Americans better than General Gage?

On paper his war record was more than satisfactory. He had acquitted himself well when 30,000 men fell on May 11, 1745, at the British defeat at Fontenoy, one of the bloodiest conflicts in the eighteenth century. He was also present at the terrible slaughter of the Highland clans at Culloden, Scotland, on April 27, 1746, when the power of the clans was shattered and the dreams of the Catholic royal pretender, Bonnie Prince Charlie, were ended forever.

On July 9, 1755, with Braddock during the French ambush near Fort Pitt (later Pittsburgh), he conducted a rearguard action that enabled the few survivors—including George Washington—to escape. He then raised a unit based on the American Rangers for forest fighting. Instead, he was ordered to lead them on a frontal assault against Fort Ticonderoga that was doomed to failure because of the fort's impenetrable abattis, and his brave unit suffered 1,600 casualties, a disaster by any measure. Gage was himself once again wounded. After the war, Gage remained in America, and in 1763 he replaced Lord Jeffrey Amherst as commander in chief of British forces in America.

There were but a few shadows on his bright résumé. Although Gage himself had become a committed Anglican with an active dislike of Catholics, he was descended from an obstreperous Anglo-Catholic family that had always contrived to be on the wrong side of every historic confrontation. Also, his detractors did not fail to point at his record in Canada; in 1759, when Wolfe's campaign won Quebec, Gage's corresponding campaign against Montreal failed in a fiasco.

Gage was "workmanlike, careful, unimaginative." He was a soldier who, dedicated to spit-and-polish discipline, was known to plan meticulously to avoid making mistakes. Just the kind of methodical no-nonsense officer that King George III liked.

The portraitist John Singleton Copley captured Gage's vigorous, handsome face with its high brow and determined mouth and jaw. In a leadership posture, the general folds his right arm across his chest and points to a battle in his background and to victory. As a depiction of the general, friends said, the portrait offered an uncanny likeness. As a predictor of the general's future, it was a failure.

Gage also had a vested interest in putting down the rebels. He had become a man of property. He owned 18,000 acres of land in New York, more in Canada, plus a plantation in the West Indies worth £600 a year. And like most of the educated, well-to-do people in the colonies, he had a profound distrust of the common man's drive for democratic government. He considered it one of the most dangerous trends in America.

The general was a likable, honorable, kindly man, and he did not realize how Americanized he had become until June 1773, when for the first time in nearly twenty years, he and his wife visited London. He was completely repelled by the rampant corruption and by his government's shocking incompetence—seats in Parliament openly purchased with bribery were a way of business in every government bureau. His wife detested England and made very clear her wishes to return home.

In February 1774, news of the Boston Tea Party gave the Gages the chance. Summoned by the king, General Gage stood in George III's office chambers in Saint James Palace in London and earnestly, convincingly, promised the king he could take care of the situation, assuring His Majesty that the problem in Massachusetts was caused by a mere handful of tavern toughs under the leadership of Samuel Adams and John Hancock. With the help of a small unit of British regulars, and with the loyal support of most Massachusetts colonists, he, Thomas Gage, was just the man to dust up those damned rebels.

That advice to the king was Gage's Rubicon. It not only earned Gage his command but also had a profound influence on the king's cabinet. Gage's comments had been translated by Lord North into direct action: the infamous Coercive Acts, which, by closing the port of Boston, were expected to recoup the cost of the tea and bring the colonials to heel.

But even before reaching Boston, Gage had begun making mistakes.

First, he made the very poor choice of the 29th Foot to close the port. The 29th Foot was scorned throughout the colonies for its history of violence against civilians, its lack of discipline, and its sullen, angry officers. In Boston in particular this was an extremely ill-advised selection, for the 29th Foot was still remembered with flashing hatred for its involvement in the Boston Massacre of 1765. In reply, the regiment, still feeling it had been set up by Samuel Adams—which was probably true—bore a burning grudge against Boston.

To compound that mistake, and against the advice of other generals, Gage decided not to post the 29th Foot to Castle William, a fortified island off Boston's shore, but to quarter it among the people in Boston—who were known to spit when they said the word *twenty-ninth.* By using a different British unit, Gage might have written a different history in that city.

On June 1, 1774, with all of Europe—including a smirking

French king in Versailles—watching, General Gage closed the port, locked a naval blockade around it, and strangely, expected the starving colonists to come to heel submissively, waving in their hands repayment for the tea. He waited—and waited—for surrender until Lexington. Vacillating day after day, remaining immobile inside the city, he let ten months pass. His troops began to grumble about him. Behind his back, the malevolent whispered he was henpecked.

After three of those ten months, on the morning of September 1, 1774—General Gage realized how much he had overpromised his king. During that morning, Gage successfully snatched 250 half barrels of gunpowder from the rebels. But the next morning he found 20,000 armed men heading for Boston, all believing false rumors that British ships were shelling Boston and that people had been killed. Awed by the fury of the colonists and their huge number—possibly even double 20,000, and surely three times larger than his army of 3,000 could deal with—Gage had learned a frightening lesson. He promptly canceled another strike, planned against Worcester—forty miles inland—and thereafter tried to avoid provocation. He probably realized that the 20,000 people who had turned out to oppose him could not all be tavern toughs.

Gage must have understood that the opposition to the Crown was pervasive, determined, and, most shocking, prepared for a relentless battle. As a consequence of the September confrontation, he changed his mind about the colonies and changed his estimates, and now urged the Crown to rescind the Coercive Acts, including the closing of Boston, until the king sent him an army of 20,000.

The king angrily dismissed the proposal. He didn't have the extra men to spare. After waiting six more months for some action, the king's cabinet ordered Gage to take action. Left to his own devices, General Gage would never have made that march to Concord. It was the strategy of an impatient king who'd grown weary of waiting and demanded a victory against the damned rebels.

Gage could pull out his direct orders from the king—the letter from Lord Dartmouth that General Gage had received just five days before he acted, assuring him in every line that he had an easy task before him. Lord Dartmouth rejected Gage's estimate that it would take 20,000 men to pacify New England. Twenty thousand men? Gage must now have wished he had asked for 40,000. The king's authority did not extend six feet outside Boston.

If Gage needed more men, he was told, he should raise a corps of infantry from "friends of government in New England." From his windows in Boston, Gage could look out on all the friends of government there were—all huddled in their threadbare Loyalist community.

George III's majesty was trotted out: "The King's dignity, and the honor and safety of the Empire, require that, in such a situation, force should be repelled with force."

To the king's ministers the uprising seemed like a typically routine matter. They had suppressed many insurrections in Ireland, Scotland, the colonies, and even England itself. Riots had become a common factor in British life. In recent years some 160 major uprisings had occurred—in Devon and Cornwall; Penryn, Truro, and Falmouth; Whitehaven; and Romney Marsh. Ireland was constantly festering. Boston rioting was familiar territory to the right troops—under the right commander.

Now, seated at his desk, General Gage looked down on a blank piece of paper. What could he say to George III? How could he explain?

Even before the battle had ended, the general had easily discovered what went wrong and how it had gone wrong. He could see clearly that his secret operation had been betrayed and botched from the beginning.

To maintain secrecy, Gage had his troops assemble on a deserted part of Boston's beach facing Back Bay, in an isolated area of the city, late at night, and in total darkness. But because of the secrecy, from the beginning there had been confusion and disorder. No one seemed to be in charge or know what the orders were. There weren't enough boats—possibly a bit of malice by Admiral Graves, who hated General Gage—so several trips had to be made and the boats were so packed the troops had to stand during the entire crossing.

Some of the blame had to go to the nasty-tempered Admiral Graves for his very public display of launching the longboats that were to carry the soldiers, then tying them under the sterns of the two British warships, where they floated at their moorings for days before they were needed.

On the other side of Back Bay, at Lechmere Point, in Cambridge, the landing site was so poorly chosen that the boats ran aground and the men had to step out of the boats into swampy water up to their knees before they could struggle onto the path. It was midnight. It was chilly. The men were wet. The crossing had taken two hours.

Precious time lost. The whole operation, barely under way, was already well behind schedule.

Then slow, ponderously heavy Lieutenant Colonel Francis Smith fussed in the darkness to assemble the troops in a certain order. More time lost. When they finally set out they found themselves slipping and sliding on marshy soil, then had to wade another river, some in water up to their waists, to gain the main road. They were wetter. And colder. And madder.

Smith then stopped to wait for the supply wagons—another hour's delay that provided the troops with foodstuff made from naval stores so full of weevils that most of the troops promptly threw it away.

In all it was a full four hours after leaving the barracks that the march finally began. The men had been turned out of their beds shortly after they'd gone to sleep, which meant that they had not had any proper rest since the night before. They had exchanged comfortable beds for hundred-pound packs. Not told where they were going or why or for how long, when they would eat, sleep, rest, or return, they had no idea whether they were going on an exercise or into battle.

And as they moved through the night, they were unaware of the enormous activity that was going on all around them, their march anything but secret. In East Cambridge, they passed by the house of Samuel Tufts, who was inside with his slave, pouring lead into bullet molds. At Foot of Rocks, they passed another house in which a man and his wife were busy melting their pewter dishes to pour into bullet molds. Everywhere in dozens of towns and hamlets, colonists were preparing for battle.

The British troops encountered an alarming number of people on the roadway and collared them to prevent any alarm from being spread. These captives were made to walk along with the army. But these efforts were useless. By this time they could hear signal guns and ringing bells from villages all around them. It didn't take a general to interpret them: Defend your towns and your families.

The story of the relief column under Lord Percy was a story of equal ineptitude. Gage's message to brigade major Captain Thomas Moncrieffe, ordering him to turn out his troops at four A.M., was delivered to his room while he was out, and when he returned, he failed to see it. He and his entire unit were sound asleep at four A.M. And General Gage's orders to the Royal Marines were sent to Major

John Pitcairn, who at that time was on the Concord road about to fight for his life. Those orders also went unopened. As a result, the 1st Brigade, under Lord Percy's command, wasn't ready to march until 7:30 A.M. And by the time the marines were assembled it was 8:30—a delay of some five crucial hours.

In the aftermath of the battle, blame for the disaster was being doled out in heavy portions from a large plate. The British had never for one instant dreamed that they could have been humbled in such a shattering experience; days later, they were still completely stunned.

Lieutenant Barker of the King's Own held Colonel Francis Smith to be responsible. In his diary, he wrote: "Had we not idled away three hours on Cambridge marsh waiting for the provisions that were not wanted, we should have had no interruption at Lexington." And he blamed Smith's ponderous slowness for failing to prevent the fighting at the North Bridge in Concord. "Being a very fat, heavy man," the angry young officer wrote, "he would not have reached the bridge itself in half an hour though it was not half a mile." He growled in conclusion: "Thus ended this expedition, which from the beginning to end was as ill planned and ill executed as it was possible to be." While much of his criticism was aimed directly at that "very fat heavy man," Barker joined with many voices in also criticizing General Gage's planning and execution.

Hugh, Lord Percy, had his own view.

A Whig favorably disposed toward the colonists when he arrived, Lord Percy, after months of witnessing the ruthless tactics of Sam Adams and dining at John Hancock's mansion on several occasions, had concluded: "The people here are a set of sly, artful, hypocritical rascals, cruel, & cowards. I must own I cannot but despise them completely." Now reassessing his scathing contempt for the American "cowards," Lord Percy wrote: "I never believed, I confess, that they would have attacked the King's troops, or have had the perseverance I found in them yesterday."

General Gage himself couldn't believe what had happened. These were not the American farmers he thought he knew so well: "In all their wars against the French they never showed so much conduct, attention and perseverance as they do now." He added, "The Rebels are not the despicable rabble too many have supposed them to be." One of the biggest "supposers" had been himself.

He was also echoing General Braddock, who in 1754 had condemned the colonials as cowards. So had the late General Wolfe, who in 1760 won the battle of Quebec and died on the battlefield. "The Americans," said Wolfe, "in general are the dirtiest, the most contemptible, cowardly dogs you can conceive. There is no depending on 'em in action. They fall down dead in their own dirt and desert by battalions, officers and all." General Amherst had scorned them as well. And Major Pitcairn himself had marched off that night for Concord believing scornfully: ". . . if I draw my sword but half out of my scabbard, the whole banditti of Massachusetts will run away."

Pitcairn had a lot of company in England. Even as late as March, only a month earlier, the earl of Sandwich, first lord of the admiralty, was assuring everyone in England who would listen that the Americans were "raw, undisciplined, cowardly men. . . . Believe me, my Lords, the very sound of a cannon would carry them off . . . as fast as their feet could carry them." This sentiment was echoed many times over by Richard Rugby, paymaster of the forces, General Sir William Howe, former royal governor Hutchinson, Lieutenant Governor Oliver, and many others.

No one in the cabinet—and few British authorities in America—believed that the Americans would fight. And when they did, the British were not prepared for the ferocity of the battle. Nor were they at all ready for the war once the Americans brought it to them.

But General Gage was not going to put any of this on paper.

In certainly the single most glaring understatement of the Revolution, General Gage dipped his quill into the ink and began his official report to Lord Barrington, secretary of war in London. Mindful that he was in effect writing the report to his king, in a flowing hand he announced the beginning of the American Revolution with a few dozen lines:

> Boston April 22, 1775:
>
> . . . I have now nothing to trouble your Lordship with, but of an affair what happened here on the 19th instant. I have intelligence of a large quantity of military stores being collected at Concord, for the avowed purpose of supplying a body of troops to act in opposition to his Majesty's Government, I gott the Grenadiers and Light Infantry out of town, under the command of Lieutenant Colonel Smith of the 10th Regiment, and Major Pitcairne of the Marines, with as much secrecy as possible, on the 18th at night, and with orders to destroy the said

> military stores, and supported them the next morning by eight companys of the 4th, the same number of the 23d, 47th and Marines under the command of Lord Percy. It appears from the firing of alarm guns and ringing of bells that the march of Lieutenant Colonel Smith was discovered, and he was opposed by a body of men within six miles of Concord, some few of whom first began to fire upon his advanced companys, which brought on a fire from the troops that dispersed the body opposed to them, and they proceeded to Concord, where they destroyed all the military stores they could find.
>
> On the return of the troops, they were attacked from all quarters where any cover was to be found, from whence it was practicable to annoy them, and they were so fatigued with their march that it was with difficulty they could keep out their flanking partys to remove the enemy at a distance, so that they were at length a good deal pressed. Lord Percy then arrived oppertunely to their assistance with his brigade and two pieces of cannon. And notwithstanding a continual skirmish for the space of fifteen miles, receiving fire from every hill, fence, house, barn, etc., His Lordship kept the enemy off and brought the troops to Charles Town, from whence they were ferryed over to Boston.
>
> Too much praise cannot be given to Lord Percy for his remarkable activity and conduct during the whole day. Lieutenant Col. Smith and Major Pitcairne did everything men could do, as did all the officers in general, and the men behaved with their usual intrepidity.

The general waited until a later report to tell the king that he had lost about 10 percent of his entire force in Boston.

After writing this report, General Gage turned to his most pressing personal problem: the person in his inner sanctum who had betrayed the Concord plan to the rebels. The general had told his plan to only two people. One was Lord Percy. And the other—he would deal with that person at once.

Margaret Kemble Gage, wife to General Gage, had watched the returning troops with mixed emotions. She was an American, but two of her brothers served in the British army, both under Gage—one of them as Gage's confidential secretary until his recent death from camp fever in Boston. In the tug of war among her loyalties to her king, her husband, and her American countrymen, she had faced a great moral dilemma. In fact, recently she had declared that "she hoped her husband would never be the instrument of sacrificing the lives of her countrymen."

Then she had made her choice.

As she looked out on the overcrowded army camp and the dozens of wounded, she knew her husband was struggling with the problem of betrayal. And he was looking right at her.

On Tuesday evening, April 18, when the troops were being mustered in the dark on the shore of Boston's Back Bay, Margaret Gage apparently met in secret with Dr. Joseph Warren, one of the leading radicals of Massachusetts, and revealed the details of her husband's military plan against Concord.

Once the battle was over and Gage became convinced of the identity of his betrayer, he shunned his wife. Margaret Gage was put on board a ship called *Charming Nancy* and was sent to Britain, a stranger in a country she disliked, away from her family and homeland, while the general stayed with his army in Boston. The marriage, never to be repaired, was effectively ended.

No decisive evidence has ever appeared naming Margaret Gage as Dr. Warren's unimpeachable source. Dr. Warren died at Bunker Hill without revealing the name. But General Gage clearly believed his wife guilty.

In the Revere home on Charles Street in Boston on Sunday, April 23, Rachel Revere, aware that her husband would not return soon, decided to send him some money so he wouldn't be dependent on charity. She located Dr. Benjamin Church, a leading patriot, who was in considerable danger visiting the city, and through him, sent her husband the extraordinarily large sum of £125, along with a tender note. Church returned to Watertown that same Sunday. He had been gone for nearly two days. He told Paul Revere he had been arrested and put in jail. He thereupon wrote a narrative of the battle of Lexington and Concord that so emphatically put the British in the wrong it would be widely reprinted, increasing Church's already high standing as a patriot.

He did not give Revere the note from his wife, Rachel, or the money.

Unable to return safely to Boston and out of funds, with only the clothes on his back, Revere himself soon became threadbare as he continued his labors for the cause. He acquired an odor so gamy that his associates tried to conduct all meetings with him out of doors.

Dr. Church made several other forays into Boston. And sometime that summer, during one of his visits, his mistress, whom he occasionally used as a messenger, became pregnant. It was the beginning of a military disaster and political scandal.

On Sunday, April 23, the indefatigable Dr. Joseph Warren of the Boston Committee of Safety began the arduous preparation of the official report on the battle of Lexington and Concord.

Unlike General Gage, who saw his report as a terse, unemotional summary of events to be read by the king's cabinet, Dr. Warren wanted a larger audience—the entire world. He saw this battle as one of ideas. He was going to put the British army on trial and, with overwhelming evidence and highly biased purple prose, expose that army as a band of monsters.

The farmers had won the military battle. Now the committee would try to win the battle for the mind.

Dr. Warren had his five score eyewitness accounts, but a massive effort was still necessary to coordinate all the documents, set them in type, and then print them. In addition, the question in front of the committee was how to write the document, wringing the last bit of propaganda value out of it, and still finish it in time to be first in London. Dr. Warren knew that General Gage was busy writing his report. Could the committee's report get to London ahead of his?

Dr. Warren prepared his preface, which throughout accused the regulars of attacking peaceable militia, of firing first, and then finished with a rousing coda: "The devastation, committed by the British troops on their retreat, the whole of the way from Concord to Charlestown, is almost beyond description; such as plundering and burning of dwelling-houses and other buildings, driving into the street women in child-bed, killing old men in their houses unarmed. Such scenes of desolation would be a reproach to the perpetrators, even if committed by the most barbarous nations, how much more when done by Britons famed for humanity and tenderness: And all this because these colonies will not submit to the iron yoke of arbitrary power."

In towns up and down the road to Concord, the justices of the peace had taken their depositions. One of the most telling was that of the wounded and captured Lieutenant Edward Gould, who had managed to drive a chaise through the many ambuscades that lay along the route to Boston to give Lord Percy the first news of the military disaster: "I, Edward Thornton Gould, of his majesty's own regiment of foot, being of lawful age, do testify and declare, that on the evening of the eighteenth instant, under the order of General Gage, I embarked with the light infantry and grenadiers of the line, commanded by Colonel Smith . . ."

Gould stated that he did not know who had fired first at Lexington. As for Concord, "the provincials came down upon us; upon which we engaged and gave the first fire." He then assured the people of Britain: "I myself was wounded at the attack of the bridge, and am now treated with the greatest humanity, and taken all possible care of, by the provincials at Medford."

In contrast to his humane treatment, other affidavits accused the British of one atrocity after another.

Benjamin Cooper and Rachel Cooper, of the town of Cambridge, where some of the fiercest fighting took place, declared that more than 100 bullets had been fired into their house and that two "aged gentlemen all unarmed" were immediately, most barbarously and inhumanely murdered by "them."

Regular James Marr, under orders from George Hutchinson, adjutant of the fourth regiment of the regular troops stationed in Boston, stated that at the North Bridge "a number of regular troops first fired . . ."

Hannah Adams averred that having just delivered a baby she found three soldiers in her bedroom: ". . . one . . . opened my curtains with his bayonet fixed, pointing the same to my breast. I immediately cried out, 'for the Lord's sake do not kill me;' he replied, 'damn you.' One that stood near said, 'we will not hurt the woman, if she will go out of the house, but we will surely burn it.' I immediately arose, threw a blanket over me, went out and crawled into a corn-house near the door, with my infant in my arms, where I remained until they were gone. They immediately set the house on fire, in which I had left five children, and no other person; but the fire was happily extinguished . . ."

The names and their statements came in a steady procession—Simon Winship, John Robbins, Benjamin Tidd, Joseph Abbot, Levi Mead, and Levi Harrington; dozens more were collected and sorted. In workaday English, their voices all put the British army in the worst light.

It was decided that Isaiah Thomas, his press now set up in Worcester, would set the type and publish. With the Gage report about to sail for London, the pressure on Thomas was tremendous. He had many pages of affidavits to typeset, days of work. Yet Dr. Warren was determined to be first.

Newspapers were now rushing horror stories into print with pens dipped in blood. In nearby Salem, Massachusetts, the local newspaper declared that the British troops had "the ferocity of a mad wild beast, . . . killing geese, hogs, cattle, and every living creature they came across . . . barbarians killed the women of the house and all the children in cold blood, and then set the house on fire . . . women in child-bed were driven by the soldiery naked into the street; old men, peaceably in their house, were shot dead . . . children were clubbed by muskets . . . depredations, ruins and butcheries hardly to be matched by the armies of any civilized nation on the globe." Even the Iroquois, it wrote, would "blush at such horrid murder, and worse than brutal rage," disregarding "the cries of the wounded, killing them without mercy, and mangling their bodies in the most shocking manner." The newspapers were putting America on an emotional roller coaster.

By Sunday, April 23, Judge Samuel Curwen of the admiralty court in Boston, one of Samuel Adams's most bitterly hated enemies, after being mobbed had had enough. He was leaving the country. Embarking on a ship bound for Philadelphia, he planned to sail to England on the next available passage. He was leaving his home, his career, his lofty position in colonial officialdom, his friends, and, he discovered with delight, even his wife. She had a terror of the sea and, after furious quarrels with her husband, had refused to go to England with him, throwing at his departing back the word *runaway*. They parted in bitterness and never reconciled, remaining enemies until death did them part.

In his will, Curwen left instruction that he was never to be buried next to his wife because it would derange him on Judgment Day to find her lying beside him.

Still "numb with horror and foreboding" from the battle of Lexington and Concord, and still too weak to mount a horse, John Adams set out on Sunday, April 23, from Braintree for the Continental Congress in Philadelphia. He rode in his father-in-law's horse-drawn sulky, passing crowds of armed farmers walking to Boston.

Having failed to meet Samuel Adams and John Hancock in

Worcester as planned due to his illness, John Adams had revised his travel schedule, determined to join his cousin and John Hancock in New Haven, Connecticut.

As he rode south, John Adams still faced grave personal danger: for using his legal talents to aid Sam Adams in what the British construed as high treason, he believed there was a noose dangling over his head, and he faced the prospect of being captured at any turning of the road. The previous December he had written: "We are, in this province, Sir, at the brink of a civil war. We have no council, no house, no legislative, no executive. Not a court of justice has sat since the month of September. Not a debt can be recovered, nor a trespass redressed, nor a criminal of any kind brought to punishment. . . ." But he was convinced the colonists would win: "New England alone has two hundred thousand fighting men, and all in a militia, established by law; not exact soldiers, but all used to arms."

As cold-blooded as his cousin Sam, and as deeply committed, John Adams said: "I would have hanged my own brother if he took part with our enemy in this contest."

On Sunday, April 23, as Israel Bissel was pressing toward his explosive reception in New York, John Hancock's carriage, drawn by four spanking horses, its trunk from Buckman's Tavern now firmly strapped to its rack, was moving furtively through the beautiful springtime of Massachusetts toward Worcester.

Hancock and Sam Adams had spent two days in the village of Billerica, where one piece of major personal business had occurred. Hancock, at last, after four years of his aunt's pursuit, had agreed to set the date: He and Dorothy Quincy would tie the knot in August, during the recess of the Congress.

On the twenty-fourth the two men arrived in Worcester, forty miles outside Boston, and feeling somewhat safer, they waited in that town to meet John Adams, Robert Treat Paine, and Thomas Cushing, the other delegates to the Second Continental Congress in Philadelphia. But John Adams, just recovering from brain fever, had already altered his travel plans.

Hancock was firing off one message after another to the Committee of Safety, demanding news, spewing advice, and insisting that the troops attack Boston. Most of all he demanded mounted escorts in full livery. "We travel like deserters," he complained, "which I will not submit to."

While lying in wait in Worcester, Hancock and Sam Adams met with Isaiah Thomas, who was spluttering with urgency. He had his press set up in Bigelow's basement, was pegging type as fast as his fingers allowed, and, now, sitting on the biggest story of the century, the biggest propaganda weapon in the history of the country, he had no paper for the next, the most historic, issue of the *Massachusetts Spy*. Hancock agreed that the need for paper was indeed urgent and sent authority to the Committee of Safety to give Thomas the paper he needed.

On April 27, John Hancock and Sam Adams decided to meet the other congressional delegates from Massachusetts in New Haven, Connecticut, and once more took to the road. No two more improbable people ever shared a horse-drawn carriage than Hancock and Adams.

Hancock, at the age of thirty-seven, had been born parson poor and then had become royally rich. The son of an impoverished Congregationalist minister, he was from one of the best families in Boston. He had inherited the enormous sum of £70,000 from his bachelor uncle, Thomas Hancock, "one of the shrewdest money makers in the colonies," a smuggler and a supplier to the British army during the Seven Years' War. With that inheritance, Hancock himself had become one of the wealthiest men in Massachusetts—so wealthy he would later offer Congress a fully-equipped man-of-war. And like his uncle before him, he was also one of New England's leading smugglers. Wags said he had the strongest personal reason for becoming a radical and it had nothing to do with politics. A break with England would erase the impossible sum of £100,000 in fines he owed the British Crown for smuggling. Vain, given to parading his wealth, John Hancock was a pushover for flattery, which Sam Adams provided from his abundant stores.

Samuel Adams, at age fifty-three, was the exact opposite of John Hancock in every way. He was born if not rich, well off, then became poor and worse, sunk in debt. He had no business sense; given Hancock's enterprises, he would have bankrupted them in short order. In fact, he had suffered a series of business failures, having lost a large sum of money his brewer father gave him to start a business, and then, after inheriting the brewery, he ran that into the ground. He capped all that by failing in the one business that should have

John Hancock

been failure proof—tax assessing. Elected a tax assessor of Boston, he was eventually in arrears for £4,000—tax money he lackadaisically failed to collect for which the courts could hold him responsible.

There was simply no way he could pay that sum or collect from the taxpayers he had let go in arrears. Only his great popularity in Boston protected him from the long reach of the law. It would have taken just a small drop in public affection for him to have been hauled into court to face charges of malfeasance in office, embezzlement, and other attendant charges—which would quickly have ruined him and possibly jailed him. But many of those who had benefited from his incompetence felt a real affection for him. Indeed, even though the stories of his tax shortages were well-known in the city, Boston reelected him tax assessor.

Hancock's reputation was equivocal. The gentlemen of Boston had little liking for Hancock, who was described as a man whose "brains were shallow and pockets deep." Because of his financing of the Boston radicals, Boston society dismissed him as "the wretched and plundered tool of the Boston rebels," "the Milch cow to the Faction," and "the poor plucked gawky." Even the radicals were not impressed. Adams himself had had little use for Hancock before 1764, when he inherited his uncle's fortune.

Samuel Adams's reputation was unequivocal. In his inner circle, among those who worked under his direction—and they included well-educated, intelligent idealistic men such as Dr. Joseph Warren—he was revered. Most people outside his own circle hated him, some with a murderous passion. He was well-known throughout the colonies, admired and emulated by some few but hated by many who regarded him as a waterfront thug, a manipulator, a political jobber, an intimidator, a blackmailer, a mob leader, a destroyer, and a newspaper propagandist of the worst sort. A criminal out of control. A terrorist and an outlaw.

In 1765, soon after Hancock inherited his uncle's fortune and was on the lookout for popularity and flattery, Samuel Adams brought him into the Patriot Club. Hancock gave the patriot cause prestige because of his wealth and his family connections. And he could also help Adams with his financial embarrassments in the tax office. In addition, Hancock had 1,000 employees, valuable votes for Adams. Hancock also paid the Club's bills. And one has to wonder how effective Sam Adams would have been without Hancock's bountiful pockets. One of the reasons Adams cultivated the very kind of man he hated in Hancock was that inheritance. "That Adams expected to do himself a good turn at the same time that he benefited the patriot cause can scarcely be doubted, but it is questionable if Hancock realized that he was expected not only to finance the Whig Party but to settle the bills of its leader."

The dissimilarities did not end there. Adams, utterly indifferent to his appearance, was a throwback to Boston's earliest days. Profoundly religious, he dreamed of returning the city to the cheerless, humorless, Bible-ridden, dour world of the Puritans, where every day was Sunday and where one either farmed or starved.

Adams hated the wealthy for conspicuously flaunting their wealth. Yet he was consorting with the wealthiest of them all, the most conspicuous of them all, the vainest of them all, fashion plate John Hancock. Hancock was fond of parading through Boston's streets in his gilt carriage, custom-made for him in England, with elegantly caparisoned outriders and gorgeously appareled footmen fore and aft, eating in the finest restaurants, entertaining in his own home the wealthiest and most powerful, serving them the best platter in the colony on the finest imported tableware and accompanied by the finest smuggled wines. John Hancock lived on the level of royalty.

During the next session of the Continental Congress, he would be seen rushing through the streets of Philadelphia with a contingent of horsemen bearing drawn sabers, fifty ahead and fifty behind his carriage. The Massachusetts militia had taken to calling him "King" Hancock.

Hancock's mansion was filled with the finest furniture, paintings, drapes, carpets, chandeliers, and decorations. Adams lived in a tumbledown house on Purchase Street facing the busy waterfront. Hancock had closets filled with the most expensive wardrobe money could buy—including a whole collection of military uniforms he planned to wear when, as he fully expected, Congress got around to making him commander in chief of the American army. Adams was so indifferent to his dress that he often looked downright scruffy.

Hancock was tall, slender, handsome in his costumes, suffering from the unwelcome consequence of high living—periodically debilitating gout. Adams was short and tubby, with a tremor of the head and hands. Hancock was a peacock, a gadabout public figure, posturing everywhere, eager to make a speech, yearning to be admired, flattered. Adams hated the limelight, did his most effective work behind the scenes, pulled strings, used other people for fronts, and was completely indifferent to credit or reputation. He never sought applause and cared not a finger's snap for the opinion of anyone.

John Hancock was after adulation. Samuel Adams was after power.

But Samuel Adams had a problem as he rolled toward Philadelphia. Clothing.

Before the First Continental Congress in Philadelphia the previous autumn, certain eyes in Boston took the measure of the seedy, threadbare Adams. Some of the wealthiest men in the colonies were going to be at that congress. Those measuring eyes decided that Samuel Adams cut too sorry a figure to be swanking with wealthy, custom-tailored Charles Carroll and John Dickinson, Virginia planters such as Washington and Peyton Randolph.

One night in his run-down little house on Boston's waterfront, Adams and his family were interrupted at their evening meal by a tailor catering to Boston's gentry. Refusing to disclose who had sent him, he measured Sam Adams from head to toe. A well-known Boston hatter then appeared to ask for Adams's hat size. Shoemakers and others concerned with eighteenth-century sartorial splendor appeared, measured, and left.

To no one's surprise, some days later a heavy trunk was placed on Sam Adams's doorstep containing a complete outfit appropriate to a conference among wealthy revolutionists—suit, shoes, silver shoe buckles, gold knee buckles, sleeve buttons, red cloak, cocked hat, even a gold-headed cane that was embossed, like the buttons, with the mark of the Sons of Liberty.

But for this visit to the Second Continental Congress, Sam Adams was once again threadbare, having abandoned his wardrobe during his flight from Parson Clarke's house in Lexington. He would now cut an Ichabod Crane figure next to the extravagantly dressed smuggler. If anything, this Second Congress was more critical than the first, and something would have to be done about Adams's rumpled, dumpy appearance.

New York

On Sunday, April 23, Israel Bissel followed the Boston Post Road to New York City, tirelessly bobbing up and down on his hasty horse and crying out the news at every crossroads in Westchester County.

This was the demesne of "Colonel" James DeLancey, sheriff of Westchester County, a man of great power and a member of the all-powerful Loyalist DeLancey clan. The colonel's reaction to Bissel's news was typical: He would form his own cavalry unit of irregulars to fight the damned "mobocracy."

In lower Westchester County and the Bronx, Bissel's stunning news quickly destroyed the peaceful sabbath for New York's land barons and their affiliated families, who lived on some of the largest estates in any of the colonies. These men were the real holders of power in New York.

In the Bronx, Bissel skirted the vast holdings of Morrisiana, the house of the Morris family, while closer to the Hudson River lay the huge Van Cortlandt estate, and near that, the DePeyster holding, which went back to the Dutch New Amsterdam days.

From there, the great landholdings stretched north up the Hudson River Valley for several hundred miles. Just a few miles up the Hudson River lay Philipsburg, a fiefdom belonging to Frederick Philipse that was so vast it extended along the river for twenty-four miles. North of that lay the estate of Beverly Robinson, son of a leading Virginia family, who had married Frederick Philipse's sister, Susannah. Their estate, Beverly, was later to play a major role in the American Revolution, serving as headquarters to General Benedict Arnold, who plotted inside the Robinson house with British Major André to surrender West Point to the British.

Above the Robinson estate lay still more wealth, still more power, in the great estates of the upriver land barons, a mixture of Old Dutch and Scotch Presbyterian dissenters, occupying one great land grant after another, stepping all the way to Albany and including the awesomely vast 160,000 acres of Livingston Manor, primary seat of the Livingstons, Scotch Presbyterian enemies of the Anglican DeLancey clan.

On Manhattan Island itself lay Bloomingdale, the estate of the bellicose Oliver DeLancey, who wielded great power from his seat on the Provincial Council. And closer to the city, in what is now the lower East Side, lay the estate of the head of the DeLancey clan—Captain James DeLancey, nephew of Oliver, cousin of the colonel in Westchester, and the single most powerful politician in New York. If America had had anything close to an aristocracy it would have been these two rival families—the Loyalist DeLanceys and the Whiggish Livingstons.

By four o'clock that Sunday afternoon, four days after the battle at Lexington and Concord, express rider Israel Bissel, having left a string of exhausted horses behind him, cantered into the sabbatical silence of New York City, the largest citadel of Loyalists in the colonies. As he cried the news of Lexington, he unwittingly thrust a lit fuse right into the touchhole of an overpacked political cannon. He was quickly surrounded by a large crowd of people who came running from every direction and then ran off in every direction as still others arrived.

Churches emptied. Bells pealed. People flocked into the streets. Spontaneous demonstrations erupted. Whigs—who just hours before had been struggling to get the merest scrap of power—paraded with joyfully thundering drums and flying colors. Lexington and Concord had changed everything. Now, at last, they had the Loyalists on the run.

The leading radical politician and street fighter, Isaac Sears—

along with two cohorts, John Lamb and Marinus Willett—was ecstatic. For the first time in his long and frustrated political life, Sears had the upper hand in the city. Determined to make the most of it, bitter from his recent jail term—Colonel DeLancey himself had arrested him for insurrection, only to have a mob free Sears—and hating the imperious Loyalists with a towering, unslakable rage, his thoughts turned to punishing his enemies.

Sears organized a mob on the spot and led them, shouting and hooting, to the city's docks. There they clambered on board two ships and confiscated supplies said to be intended for the British garrison. Then they overran a sloop ready to sail for Boston with more British military provisions, and looted it to the bilges.

That evening, the roaring Sears led the rioters by torchlight to City Hall, where they smashed the locked doors, broke into the city's arsenal, and carried off 550 muskets. Shouldering those weapons, squads of men attempted to learn military drill on the streets of the city. Self-appointed censors stopped post riders to inspect letters "for anti American [pro-Loyalist] sentiments."

The sounds of beating drums and smashing glass, the scenes of mobs clambering over ships along the piers, and the sight of throngs of people running through the streets announced that the radicals were seizing the control so long denied them. Until the authorities restored some semblance of order, every Loyalist in the city was suddenly in danger. Many of them quickly left for friendlier places.

Besides sustaining the riots, which he would keep up for days, Isaac Sears was particularly interested in hunting down three men: Reverend Dr. Myles Cooper, newspaper editor James Rivington, and Reverend Samuel Seabury. In the days that followed, Sears led his mobs in pursuit of them.

Dr. Myles Cooper was president of King's College (later Columbia University), located in lower Manhattan, not yet twenty years old and consisting of a single building and forty students, The year before when John Adams had visited King's College he had sniffed, "Not to be compared with Harvard."

Reverend Cooper, among his other quixotic gifts, had a decided knack for gathering enemies—in enormous numbers. Just being an Anglican minister in New York City in 1775 was enough to draw them. Anglicanism was the official, state-controlled religion of the British Empire, with the king himself at its head. Support for the Anglican Church came from tax money paid by all subjects in the British Empire, including unwilling dissenters. And therein lay much of the rub. In New York, relations between the Anglicans and all the other dissenting religions were murderous.

Reverend Cooper was quite aware that when he walked along the streets of Manhattan, tipping his hat affably to the world, members of dissenting religions would feel the gorge rise in their throats; sure of the rightness of his cause, every Sunday he would mount his oaken Anglican pulpit, sleek and well fed on their money, to preach against them and their fervently held beliefs.

Hewing to Anglican doctrine, he defended—demanded—the distinctions between classes, stressed obedience to church and king, and endorsed "the Crown's economic program which required submission and hard labor from the colonies for the enrichment of the mother country." Most particularly he roundly damned the civil disobedience of the radicals as a major sin.

The dissenters felt that their own money was—against their will—hiring people such as Dr. Cooper to attack their religions. But for Reverend Dr. Cooper, that was hardly enough turmoil; there was still much more of the Lord's work to be done. Reverend Cooper and a group of other Anglican divines, rushing in where few others would, had plumped to have the bishop of London appoint an American Anglican bishop as a way of undermining Congregationalist, Presbyterian, and "other dissenting denominations" in the colonies. To the other religions, which had come to America to escape Anglican dominance, this appointment could be exceeded in horror only by the appointment from Rome of an American Catholic bishop. It was a prospect that kept all the other religions enflamed and the New York political scene in continuing turmoil.

In the grinding warfare of American politics—the Stamp Act, the Townshend Acts, the Boston Massacre—appointing an American bishop was an explosive issue. The bishop of London exhibited more common sense than his American ministers—he refused to consider it. But the American Anglicans kept pressing, kept incensing without rest dissenters such as the dangerously powerful Livingstons.

This was enough to supply Reverend Cooper with enemies everywhere he turned. But his restless nature looked for still more controversy, more sinners to belabor. He was a preacher. He was also a master of the poison pen. He felt he should attack the enemies of his religion. So he sallied forth through the pages of the *New York Gazetteer,* where, week after week, knocking sconces, joyfully laying about left and right, he scorned, mocked, and ridiculed both religious and political dissenters.

By damning the radicals and calling them "sons of licentiousness, faction and confusion" and more, he earned himself the passionate hatred of the local Sons of Liberty—and in particular Isaac Sears.

So with the news from Lexington enthralling New York City, it was time for the reverend to wind up his affairs and leave the city by the shortest route. He was warned in advance of the approach of the marauding Sears mob. They were coming, he learned in vivid detail, "bent on shaving his head, cutting off his ears, slitting his nose, stripping him naked and setting him adrift." But he delayed.

On the way to visit him, the mob also delayed, stopping long enough to fill up on drink, giving the Reverend Cooper ample time to abscond. But he had become paralyzed with fear and, foolishly, he was still inside the college building, on an upper floor, when the furious mob approached and prepared to mount the stairs. Cooper was trapped inside.

Then the reverend got an incredible stroke of good luck.

Unexpectedly, a twenty-year-old student stood on the steps of the college, placing himself between the mob and the front doors. He harangued the crowd, soothed them, cajoled them, and seemed to have saved Cooper's life. His name was Alexander Hamilton. But, unable to hear what Hamilton was saying, the bumptious old divine opened the window and shouted down at the mob: "Don't believe anything Hamilton says; he's a little fool!"

Dr. Cooper, haunted for the rest of his life by the memory of that mob, promptly fled headlong to the safety of a warship in the harbor—the HMS *Kingfisher.* A week later, he continued his flight to England on the HMS *Essex,* taking the official Anglican religion with him. His enemies had finally prevailed; his days of taunting them were over.

Young Alexander Hamilton had managed to halt the mob on the school's steps because he was a fellow radical. He had joined the radical cause the previous year—1774—after he had visited Boston to observe the situation firsthand. There he met many of the rebel leaders—including Sam Adams and John Hancock—and became a convinced radical against the Crown. Returning to New York, he wrote two highly influential, anonymous pamphlets that were remarkably mature, displaying "such a power of reason and devastating sarcasm, such a knowledge of government and the English constitution that would have done honor to any man at any age."

With the news of Lexington, Hamilton realized school was over. He decided to join the American army as an artilleryman.

Isaac Sears now turned his triumphant eyes on his two other enemies, Rivington and the Reverend Seabury.

Waspish, foppish James Rivington, fifty-one-year-old Loyalist publisher of the weekly *New York Gazetteer,* was one of the most colorful and controversial newspapermen in America. He was famous for a razor tongue that could surgically remove the flesh from the most case-hardened politician. His lively, gifted, antiradical invective drew the hotly cannonading enmity of the colonial radical press everywhere. "Jemmy" Rivington was at the top of the radical's hate list.

Rivington had learned his trade from his father, a prosperous London bookseller and printer. A young man on the make, Rivington came to New York in 1760, set up shop as a publisher and bookseller, quickly prospered, and soon married the kind of money that not only enabled him to live luxuriously but also increased his social stature. Oozing success and well-being, mingling with the patrons of his bookstore, directing his printers, editing his paper, Rivington cut an extremely stylish figure in a "rich purple velvet coat, full wig and cane, and ample frills." Mixing affability with nimble intelligence, he was an entertaining, witty companion with the breeding of a London gentleman.

He soon contrived to know everyone in the colonies, either through their visits to his bookshop or his extensive correspondence. His friendship even extended to his fellow bookseller and correspondent Henry Knox, who would later publish a revolutionary newspaper in New Jersey for the American forces that eventually became the *Elizabeth Daily Journal.*

Actually, the *Gazetteer* was a fine newspaper. One of the very few pro-British papers in America, the *Gazetteer* had outstanding foreign news and very popular literary miscellany. With a level of writing well above that of other papers, the *Gazetteer* was probably the best-edited, most informative newspaper in the colonies—Whig or Loyalist—and typographically, its handsome, tasteful design had few equals. In time it became one of the most widely circulated papers in the colonies.

Two years before the battle at Lexington almost to the day on April 22, 1773, Rivington launched his *New York Gazetteer* with such instant success that another paper in the city, the *Gazette or Weekly Post Boy,* soon folded.

Originally, Rivington sought to maintain an impartial editorial policy, accommodating the views of both radicals and loyalists. But in the turbulent 1770s, readers had no patience for reasonableness, demanding instead that their news be cut on the bias and presented with plenty of bite. Rivington wasn't pleasing anyone. In fact, the radicals in New Brunswick, New Jersey, burned him in effigy. Rivington was so incensed he instantly became an ardent foe of the radical cause.

So on November 18, 1773, safe in the British garrison town of New York City, Rivington introduced his new pro-British editorial policy and began a taunting editorial campaign against the radicals and Whigs that exposed their tactics, ridiculed their solemn pronouncements, and maintained a steady, unmerciful drumbeat of invective that appeared issue after issue.

Rivington scorned the revolutionaries as "obscure, petty fogging attorneys, bankrupt shopkeepers, outlawed smugglers, wretched banditti, the refuse and dregs of mankind." In particular he held up to ridicule Isaac Sears, calling him "a tool of the lowest order," a "political cracker," and, most infuriating, "the laughing stock of the whole town." Sears hated Rivington so passionately he vowed that at the first opportunity he would scatter Rivington's type, wreck his offices, and worse.

Despite veiled threats in Whig journals in New York, Rivington continued his relentless parade of slashing articles, many written by Dr. Cooper. In time, Rivington had as many as thirty Whig newspapers all up and down the coast baying in full furious cry after him, writhing with an ecstasy of rage at his cool invective.

For a polemicist, Rivington achieved pure nirvana: The *New York Gazetteer* became the most hated Loyalist paper in the colonies and also the most widely read up and down the coast—for a while. By October 1774 Loyalists throughout the colonies—as well as fulminating radicals—had raised his circulation to a heady 3,600, unmatched by any radical paper. In Boston, General Gage himself bought and distributed 400 copies of each issue. But by mid 1775 fewer and fewer copies of the *Gazetteer* were getting through the radical postal service. It had also become dangerous to receive the paper, and it began to fail. As Israel Bissel cantered into New York that Sunday afternoon, Rivington was facing bankruptcy and lived in increasing danger of physical assault. His safety was hanging by a thread.

When Sears finally got up enough nerve to sic a mob on Rivington's home and printery, the publisher did the wise thing. He, like Dr. Cooper, fled to the *Kingfisher*. Safely behind massive naval cannon, they together could watch the rioting mobs ranging along the waterfront.

Sears had to wait to get his revenge on Rivington. In November he swept down from Connecticut into the city leading a posse of mounted thugs with fixed bayonets to raid Rivington's newspaper. As he had vowed, Sears and his men smashed Rivington's presses and scattered his type.

Dauntless, Rivington bided his time, waiting for the British army to reoccupy New York. He had more to say about Isaac Sears.

Hamilton had words for Sears as well: "Though I am fully sensitive how dangerous and pernicious Rivington's press has been, how detestable the character of the man is in every respect, yet I cannot help disapproving and condemning this step."

The third man on Sears's list was Reverend Samuel Seabury, minister of the Anglican parish of Saint Peter's in Westchester, New York. Reverend Seabury, a stripling at the age of twenty-seven, once observed, "I find it very difficult to convince people that religion is a matter of consequence." Arguing one day with a particularly obtuse and mendacious individual, he found his tussle "the most disagreeable task I ever undertook. 'Tis like suing a beggar, or shearing a hog, or fighting a skunk."

In the feverish days of the 1770s, however, especially in the days of the First Continental Congress, Reverend Seabury, a dedicated Church of England man, felt moved to attack the stupidity and mindlessness of the radicals. Their arguments sent him into lofty heights of pen biting.

Reverend Seabury suddenly discovered an uncommon gift for political polemicism. Using the simplest of prose and always making a direct frontal attack on his opponents, he learned he could tear the hide off the most stubborn and inane arguments. He did not have any difficulty convincing people that politics was a matter of consequence.

Reverend Seabury, with pen and paper, waged a remarkable war.

When the First Continental Congress was in full chorus in Philadelphia, Sam Adams's plan for a nonimportation program caught the reverend's eye and ire. In response, Reverend Seabury brought forth the most sustained, the most wilting, attack on radicals in all of the American Revolution. His format was a series of pamphlets containing four letters from an anonymous Westchester farmer. In an age of elaborate English, Seabury's prose was shockingly simple and clear, supple and muscular, quite appropriate for a no-nonsense farmer—which the reverend also was, farming the land that came with the parish's house. Seabury used no abstruse theory, no flowery embellishments, no parade of classical learning and apt quotations, just blunt, practical, sensible cautions—with brilliant periodic rockets of invective. And that made his pamphlets devastating.

The first discussed farm matters, bushel costs and income and other practical farm economics. He demonstrated to the farmer that farmers alone would be the victims of the nonimportation agreement. Farmers without debt soon would be in debt, and farmers in debt would soon be ruined. Unable to sell their produce, they would lose their farms to foreclosure—while self-appointed busybody city radicals from self-created nonimportation committees would "search your pantries, your wives' and daughters' petticoats looking for contraband." Worst of all, if the farmers survived the embargo, afterward they would find that their markets in Ireland and England had gone glimmering, lost to farmers in other countries. Businessmen, however, he warned the farmers, would prosper through smuggling. As usual.

The second pamphlet, directed at businessmen, asked a single terrifying question: If the British win the war what will the king do to you and yours? The question was rhetorical. His readers knew quite well what the king would do. Seabury told them nonetheless. Even if you win, he warned them, a most improbable outcome, you will lose, for surely the victorious colonies would then rend each other in a terrible civil war and destroy each other. "But it is much more probable," he wrote, "that the power of British arms would prevail; and then, after the most dreadful scenes of violence and slaughter, confiscations and executions must close the horrid tragedy."

Among other things, the essays condemned the First Continental Congress for what it was—illegal and definitely not representative of the feelings of most colonists. A handful of willful men were pushing the rest of the colonists into a face-to-face and unwinable confrontation with the Crown. The essays also held up to ridicule the idea that a tea tax was a deprivation of liberty. And they cried shame down on the terrorism and beatings that the radicals such as Sam Adams employed on merchants to enforce the sagging embargo.

The letters had a stunning effect. They frightened many people. And made many think. They shook the resolve of many radicals, caused the borderliners and the fence-sitters to waver, and gave nightmares to many who had not thought clearly about the consequences of failure.

Widely distributed, widely read, the letters so infuriated the Sons of Liberty that they tied one to a tree and literally tarred and feathered it in lieu of its unknown author.

In his church when news of Lexington reached him, Seabury read the handwriting on the wall. Since some people had guessed that he was the author of the letters, he decided it would be prudent to absent himself from the precincts of Westchester until affairs sorted themselves out. Unlike Reverend Cooper, Reverend Seabury did not wait for a Sears mob to arrive with the tar and feathers.

If he had had the nerve—he didn't—Isaac Sears would have led his mob up to Bloomingdale, the estate of Oliver DeLancey.

Oliver DeLancey never left anyone within shouting distance in doubt for long about his opinion on any subject. And though an unshakable Loyalist, he was also known far and wide—even in London—for his unedited, intemperate diatribes against Parliament's attempts to tax the colonies. When he heard the news from Massachusetts, he suspected immediately and loudly that the Lexington affair was an exquisitely timed move by Sam Adams to push the upcoming Second Congress in Philadelphia into revolution. Regarding Sam Adams as a dangerous radical who long ago should have been hung from the nearest liberty pole, DeLancey realized that, with the affray at Lexington, colonial politics had shaded from street violence into revolution. With the howling mobs parading in the downtown area, he was no longer safe in his own city. It was time to prepare for war. And he had only hours to act.

DeLancey's power was based on familiar components—the right connections and the right antecedents. First of all, he was staunchly Church of England Episcopalian—the only thing for a royalist to be. Second, he had been educated in England at Eton and Cambridge, which gave him many invaluable contacts in London—in Parliament and the royal court—and also put him on an equal social footing with the many British government officials and army officers in the colonies, among whom he had many friends. His roots inside the British army went even deeper: After college, he had dutifully taken an army commission, which gained him command of a British army regiment during the French and Indian War. His military record was twice officially commended by the New York Assembly.

There were powerful connections through his children also. His son, Oliver Junior, was a career officer in the British army. During the American Revolution, he would rise to the rank of adjutant general of the British army in America, replacing Major André, who was executed by Washington as a spy. Oliver Junior, went on to be among England's top-ranking officers during the War of 1812.

Another son, Stephen, held the important position of clerk of the city of Albany and became famous in 1776 for singing "God Save the King" with a group of roisterers, for which he was thrown in jail. He emerged from prison, sobered and angry, to become commandant of the Loyalist 1st Battalion of New Jersey Volunteers, and participated in some of the most savage fighting between rebels and Loyalists in the war.

Back home at the end of the French and Indian War, Oliver DeLancey Sr., turning to politics, soon rose to a powerful seat on the Provincial Council.

With the right marriage—to Phila Franks—he made a whole new set of influential connections from Canada to Georgia. His brother-in-law was David Franks, a fifty-five-year-old Loyalist merchant born in New York, the fourth generation of a Jewish merchant family with strong trade connections in Britain, the colonies, and Canada. Franks lived in Philadelphia with his wife, who was a Christian.

Thus, Oliver DeLancey was a man of considerable political power in his own right. And he enhanced that power greatly when he joined his post in New York's Provincial Council with the power base of his nephew, James DeLancey Jr., who controlled the Colonial Assembly. The result was a political fortress that seemed to their opponents to be well-nigh impregnable—until Lexington and Concord.

But Oliver DeLancey knew on that sunny Sunday in April that his great political power had just evanesced in a puff of smoke. As he had expected, the morning after Bissel's ride into New York the "Rebel Committee & Provincial Congress" summoned him to appear before it. Rather than submit to that star chamber indignity, De-

Lancey betook himself to the seventy-gun British warship HMS *Asia,* anchored in the harbor, and remained on board until General Howe landed his troops on Staten Island. There DeLancey laid plans for war, forming three Loyalist army units. DeLancey was to fight on the Brooklyn Heights and in Flatbush and, as brigadier general, would become the ranking military officer in the Loyalist corps.

From his forty-acre estate on the Brooklyn Heights, Philip Livingston was preparing to travel to Philadelphia and the opening session of the Second Continental Congress. If anyone had told him at that moment that he would sign the Declaration of Independence the following year, he would have believed he was talking to a madman. Livingston could see for himself the crowds cavorting with torches through the streets of Manhattan, across the East River, making the idea of a break with England too improbable to conceive. The prospect was terrifying. He had written just the year before, "If England should turn us adrift, we should instantly go to civil Wars among ourselves." The "Leveling Spirit" was as much to be feared as the civil war.

At the age of fifty-nine, a Yale man, Philip was a prosperous city merchant with a thriving import business. He had added immensely to his great inherited wealth during the French and Indian War when, as a privateer, he snatched a fortune in enemy shipping from the high seas.

The evolution of his political philosophy mirrored the events of his times. He began as an abiding conservative who flatly opposed the idea of American independence, the secret ballot, and the popular vote. But as the colonies' relationship with England turned turbulent, he became increasingly hostile toward the mother country. As early as 1754, while he was serving as an alderman in New York City, he developed a biting opposition to the Anglican ruling class of the colony, especially the high-handed, autocratic tactics of James DeLancey Sr.

Elected to the assembly in Albany, he was soon fighting the royal governor, opposing British treatment of the colonies, and gradually, finally, becoming an anti-royalist Whig. He had attended the Stamp Act Congress in 1765 and, moving ever farther away from George III, had actively supported its anti-Parliament resolutions. The rift between him and his conservative heritage had widened even more when he attended Philadelphia's First Continental Congress.

But Philip Livingston was no unbuttoned rabble-rouser, and at every turn he firmly opposed the riotous behavior in New York City of the radical McDougall-Sears combine and the Sons of Liberty. Eventually, he and many of his class would feel rootless in the new democratic country that would emerge from the Revolution. There was in them, underneath all, a strata of strong distrust of the common man's desire to run his own affairs. Like other members of his family and of his class, Livingston believed that "only men of quality are qualified to lead." In the end, the long philosophical journey he had taken from conservative to democrat did not go far enough, and he would find himself on the very fringe of the new egalitarian society that toddled out of the Revolution.

As he roamed around the city with his mob, Isaac Sears, most of all, wanted to get his tarry fingers on Oliver DeLancey's nephew James DeLancey Jr., the son of the former lieutenant governor of New York, the elder James DeLancey. Captain James DeLancey, an absolute autocrat at the age of forty-three, and head of the New York Assembly, was in the prime of his political life.

James Junior felt that New York City was his town. He had been born there in 1732 in the DeLancey mansion, which would later be converted into Fraunces Tavern, and—like most of his male relatives, including his father and his uncle Oliver—he had been educated in England, at Eton and Cambridge. He also, like them, was a commissioned officer in the British army, with the rank of captain. During the French and Indian War he had campaigned with distinction at Ticonderoga and Fort Niagara. Upon the death of his father in 1760, when James Junior was only twenty-eight years old, his inheritance made him one of the richest men in the colonies.

In his youth, James Junior had been addicted to horse racing, cockfighting, and women. After his visit to Philadelphia with twenty-three "gay blades," that city denounced him a rake and a wastrel and, fearing that DeLancey and his cohorts would "debase our Manners," threatened to shut the city gates against his future visits.

But DeLancey had restrained himself long enough in Philadelphia to woo and win Margaret Allen, daughter of Pennsylvania's chief justice and granddaughter of Andrew Hamilton, who, in his famous defense of Peter Zenger, won the cornestone victory for freedom of the press in America.

Passionately interested in horsemanship, the captain imported the first English racehorses in the colonies and proceeded to develop the largest American stud and racing stables of thoroughbreds, ultimately developing the breed that would become the Morgan.

Captain James was undoubtedly the most gifted machine politician in New York and one of the very few politicians who were matches for Sam Adams. In fact, Captain James DeLancey was a mirror image of Sam Adams and, independently of Adams but by using the same techniques, had for years skillfully kept Adams's supporters in New York—particularly Isaac Sears and Alexander McDougall—from accumulating any meaningful power.

Just as Adams controlled the legislature in Massachusetts, Captain James DeLancey controlled the legislature in New York, through which he pushed antiradical bills at will. If Adams operated out of the Green Dragon Tavern, Captain James operated from the Fraunces Tavern, known then as the Queen's Head. And when Sam Adams began the formation of the intemperate Sons of Liberty clubs throughout the colonies, James DeLancey defused the move in New York City by forming his own Sons of Liberty, stuffing it with conservatives and rigidly excluding all radicals—an anti-Sons-of-Liberty Sons of Liberty.

Also like Adams, the captain recognized early the awesome power of newspapers and staunchly promoted Tory editors. In fact, New York City boasted more Tory newspapers than all the other colonies combined. And among them thrived the most brilliant and most widely hated Tory paper in all the colonies, Rivington's *Gazetteer.*

Captain James DeLancey also drew great power through church support. Matching Adams's "Black Regiment" of Congregational preachers, DeLancey had Church of England ministers such as the Reverends Cooper and Seabury rocking their pulpits every Sunday against "mobocracy." Furthermore, matching Adams's use and abuse of the law, DeLancey also used and abused the law. He hounded New York radicals through the courts and was not above tossing them into jail—which he did to both McDougall and Sears.

But there the similarity with Sam Adams ended. For DeLancey did not employ mobs and thugs, did not endorse tarring and feathering. As a wealthy patrician, Captain James believed, as did all the other members of his class, that it was the duty of the upper classes to think and act for the incompetent masses and control all political power. As an Anglican, he believed that all persons should be required to attend the state-controlled Anglican services and that all other churches should be shut down.

The captain, in sum, was everything that Sam Adams hated.

There was one other difference. For Sam Adams, the tide against British arrogance and brutal control was running at flood stage. For the captain and his patrician view of the world, time was running out. He was guarding the pass at Thermopylae, a holding action that would eventually fail.

The captain's career is a textbook on brilliant political oneupmanship. He fought three different opponents on three different fronts by playing one off against the other—Parliament in London, Scotch Presbyterian Livingstons in Albany, and the radical street politicians in New York City.

On the hustings, James used Sam Adams's major principle: "Put your political opponent in the wrong and keep him there." To keep the Livingstons constantly off balance, Captain James accused them of being a pack of pettifogging lawyers who were planning to grab all the property in the province for themselves and impoverish the workingman.

"Does not this Country as much swarm with them as Norway does with Rats?" he demanded. He urged the voters to kick out "these pettifogging Attornies . . . the bane of society!" and replace these vermin from the country districts with sound businessmen who had the best interests of the province at heart. Under the slogan of "No lawyers in the assembly" DeLancey won a landslide victory at the polls—a disaster for the Livingston faction.

As for controlling the workingman, Captain James kept them from getting the vote, particularly the secret ballot they hankered for. And he left the inept Sears to his own devices, for Sears could always manage to put the lower classes in the worst light by goading them into street violence—his angry tavern talk to the discontented and powerless, his bellowed threats and his manipulation of the waterfront cabals would always end up with a mob raging through the city.

After the Stamp Act riots, during which workingmen and seamen destroyed thousands of pounds of property in lower Manhattan, the captain and his entire class withdrew their support from street politicians such as Isaac Sears. The line was drawn deeply, particularly after Sears and his waterfront gangs turned their fists on New York's merchants in an attempt to make them maintain the crumbling nonimportation agreement against England. To add insult to grievous injury, the merchants showed that they were even ready to confront the street gangs with canes and clubs of their own—and actually scuffled with them to a convincing draw on Wall Street.

The most destructive challenge to the captain's power came not from the Livingstons or from Sears but from a stubborn Scotsman named Alexander McDougall. At the age of forty-three, he was the same age as Captain DeLancey, yet these two could not have been more opposite. This confrontation between Captain DeLancey and Alexander McDougall is a classic story of the politically nimble patrician confronting the flat-footed street brawler who for years relentlessly stalked him.

An immigrant Scot, Alexander McDougall was an avid student and emulator of Sam Adams, and he liked to be described as the Sam Adams of New York, which he hardly was.

By his dress and lifestyle McDougall proclaimed his modest origins. He was a boy of six when his father emigrated from the island of Islay, in the Scottish Hebrides, to New York City, where he became a milkman. The boy started working as a small child, delivering milk to the homes of the city alongside his father.

McDougall gave no hint of being a future political warrior when he went to sea as a deckhand while still a boy. By his mid-twenties, a hardened seaman with a powerful body and a roughly handsome face, McDougall had become a ship's captain. Soon afterward, he acquired his own ship, and in the next few years he went privateering in the Caribbean against Dutch shipping—first in 1757, with a six-gun ship, the *Tyger,* and a crew of fifty; then in 1759, with the nine-ton, twelve-gun sloop *Barrington,* with eighty men. He had made a fortune at sea before he was thirty. At war's end McDougall invested his capital in land and trade and money lending, factoring to both St. Croix planters and New York merchants.

His wealth made him welcome in the Livingston family's Wall Street Presbyterian Church, where he became an important financial contributor. Seeking to increase his stature, he sent his sons to Princeton's forerunner, the Presbyterian College of New Jersey, to get the education he never got. He dressed in expensive clothing, carried a snuffbox, had a saddle horse, even acquired books and comported himself like a gentleman.

But the old wealth—the powerful of the colony, the DeLanceys—didn't accept him. They disliked his garish clothes and the unlettered manners of a common seaman who preferred rum over wine. Despite his thick Scottish burr and a pronounced stutter that prevented him from becoming an effective platform speaker, they accused him of talking too much.

With no love for the English who had conquered his native Scotland, no love for the British navy which he remembered sullenly from his sailing days, no love for British trade laws which throttled his activities as a city merchant, and no love for the patrician Anglican English snobbery of those who barred his way at every turn, he was, predictably, drawn to the Whig side of politics.

Yet, raw and unlettered as he was, the bellicose McDougall could not be casually dismissed by the political powers. As a former ship's master and privateer captain, and now a money-laden trader, he had considerable stature along the waterfront, where he could muster on short notice a hard core of sailors for riots that would keep the turbulent political pot boiling. He was also highly intelligent, a keen observer, a thinker, and a dogged learner who taught himself the art of challenging authority and wresting away power in raw hunks.

Ex-army captain DeLancey soon discovered that ex-ship's captain Alexander McDougall was a very dangerous man.

What brought on the confrontation between the two men was McDougall's anonymous handbill, "To the Betrayed Inhabitants of the City and Colony of New York," attacking Captain DeLancey and the New York Assembly for accepting Parliament's Quartering Bill, which dealt with the housing of British troops in New York. Unfortunately, McDougall had been identified as the author of the piece. Under the libel laws of the time, the piece was clearly a libel against the assembly and thus McDougall was subject to a heavy fine and a stiff jail sentence.

So the captain went for McDougall. A good stiff jail sentence would help quiet him—and New York—for a long time. To make an ironclad case against McDougall, all Captain DeLancey needed was one witness to McDougall's authorship of the libelous tract. And he had that witness: none other than the printer of the piece, James Parker.

McDougall was bound over for trial by the supreme court. But then, for once, the captain's luck ran out. His prize witness unexpectedly died. And the libel case died with him. McDougall was free. And angry. A great celebrity among Whigs and patriots in every colony, he became a popular hero. The Sons of Liberty extolled his name in every meeting, the self-sacrificing paladin of the common man.

But DeLancey had done severe damage to the radicals, and he finally showed the total political weakness of McDougall when he orchestrated in the assembly the defeat of the McDougall-favored bill that would have given the power of the secret ballot to the workingman. McDougall and his cohorts were "thoroughly defeated in nearly every way," and the captain's approval rating soared. Sick of the radicals' rabble-rousing and the street violence, the patricians and the merchants lined up solidly behind him.

The fortunes of McDougall and his cohorts Sears and Lamb had sunk so low that they were unable to do anything right. The radicals even botched their version of the Boston Tea Party when "Mohawks" dumped tea from a ship into New York Harbor. But this tea was the personal property of the ship's captain, a man named Chambers, and was not Crown property. So their action failed to draw the same attack from Parliament that the Boston action did. The New York tea party barely made a ripple in the harbor. People were beginning to laugh at the radicals.

And when McDougall and Sears called for a meeting of patriots to choose delegates to the upcoming First Continental Congress, DeLancey delivered another brilliant, totally unexpected blow when he swamped the meeting with 300 conservatives, including some of the city's wealthiest merchants, who seized control of the meeting and selected a committee of fifty-one, a contingent so large and so unwieldy that McDougall was unable to wrest control of it. Worse yet, the committee chose conservative delegates—Philip Livingston, Isaac Low, James Duane, Joseph Alsop, and John Jay—thereby completely blocking McDougall, Sears, and Lamb from becoming delegates. DeLancey had beaten McDougall and the radicals once again. More scornful laughter.

After this rebuff, the city Whigs were so intimidated they did not dare talk politics in public. But Alexander McDougall, unlike Sears, had learned from every defeat and, finally getting the hang of politics, was about to make the first serious breech in the captain's fortress.

McDougall called a meeting of workingmen only—which shut out DeLancey's followers, the patricians and merchants—and guided the workingmen into passing resolutions and selecting a rival slate of five candidates that included McDougall himself. Contending that his workingman's meeting was as valid as DeLancey's committee of fifty-one, McDougall employed the agitators' principle "If you can't convince, confuse." So he went through the city, issuing contradictory claims, and by making a great show of having nine of his McDougall faction resign from the fifty-one, thoroughly confused everyone. It was all pure Sam Adams political theater, and it caused a serious gridlock.

The captain was faced with the embarrassment of sending no delegates to the very congress he had called for. To avoid this, the captain came to the table to deal with McDougall and seemed to have won another victory; McDougall didn't get a single delegate onto the list. In fact, he even agreed to support DeLancey's slate of candidates to the Continental Congress. There was only one hitch: The delegates were instructed to vote for another boycott of British goods.

It was a slight gain, but McDougall could now contend that he was in control of the political situation. After all, the captain's delegates were setting out with McDougall's instructions. And since this was the first breach in DeLancey's absolute power, McDougall had his first reason to be hopeful. But it was the shifting of the political ground, not McDougall's skill, that was working against DeLancey now. McDougall, assessing the changing political climate, now had good reason to believe he could win at last.

Sam Adams, by physically intimidating the moderates, especially Joseph Galloway, in Philadelphia, quickly gained control of the First Congress, passed the defiant and incendiary Suffolk Resolves, and called upon the colonies to accept a second nonimportation policy. Nothing in that congress pleased Captain James. He replied to Sam Adams by getting the New York Assembly to refuse to endorse any of the resolutions of the First Continental Congress, labeling them as treason and rebellion. This ripped a gaping hole in the facade of unity the congress wanted to turn to the king. But the captain was holding on to his power by his fingernails. He defeated the three pro-congress bills by a single vote, his own, 11–10. Had there been a shift of just one vote, the captain would have lost. Undaunted, the captain lifted the radicals to new Himalayas of impotent rage when he led the assembly in the drafting of a new petition to Parliament and the Crown humbly pleading for reconciliation.

The radicals were frothing at this.

Then came the crowning insult—a crushing blow to the patriot cause. The captain got the assembly to refuse to send any delegates from New York to the Second Congress. McDougall and his forces took to the hustlings, beating fists on palms and damning the DeLancey faction as traitors, despoilers of freedom, and worse—in the pay of Parliament. McDougall seemed beaten. The Tories had shown the colonies that they—not McDougall's and Sears's street gangs—still ran New York. The only response McDougall and Sears could make was to try to enforce the congressional non-importation resolution that the captain's New York Assembly refused to endorse.

In February, McDougall and Sears turned to intimidation and violence. When the ship *James* arrived from Glasgow, a waterfront mob drove it back to sea still laden with its cargo. In a sloop filled with a group of armed men, Isaac Sears prowled the bay to drive off more ships. But driving a few ships out of the harbor hardly constituted a unanimous king-defying embargo. Worse, the violence played into the hands of Captain James, who used it as a reason to fling Sears into prison.

Sears now became "the butt of every puny jester, and the laughing stock of the whole town." He indeed seemed, as Rivington had termed him, an "exploded cracker." Even his rescue from jail by a McDougall mob failed to help the radical cause. To the Loyalists, it was just another example of radical violence and lawless behavior. With great disgust, the Loyalists condemned the New York magistrates who surrendered Sears to his friends for not having "the spirit of a louse." Sears was out for revenge. But his moment hadn't come yet.

McDougall, however, now an apt student of the captain's methods, showed he was mastering his craft by making one of the smartest moves of his political career. He announced that, since the assembly refused to name delegates to the Second Continental Congress, a new committee of sixty would.

With the tenacity of a pit bull, on March 6, little more than a month before the battle at Lexington and Concord, he assembled an extralegal Committee of Inspection. This committee called for a subcommittee that would prepare a slate of deputies who in turn would be authorized to choose a list of delegates to the congress—but only after that list of deputies had been approved in an election by the city's freeholders and freemen. It was convoluted, committee-stacked-on-committee legerdemain, but it made the man in the street forget that the base of it was extralegal twaddle.

With the Liberty Boys doing a lot of pushing and shoving, and the Loyalists doing an equal amount of protesting, McDougall was able to get the subcommittee of the Committee of Inspection to select a proposed slate of eleven deputies. The city's freeholders and freemen were scheduled to vote on March 15 at City Hall to approve the slate. That slate of deputies, which McDougall fully expected to control, would then choose radical delegates for the May 10 Second Congress in Philadelphia. Sam Adams would have been impressed. McDougall was draining power from the legal government into extralegal committees and congresses. And the more the legal government remained intransigent, the more power it lost to McDougall and other radicals.

Circumstance was now playing into McDougall's hands. The king, still wholeheartedly addicted to working against his own best interests, made another major political blunder that completely undercut the Loyalist power base in New York—DeLancey's Tories.

Just prior to the March 15 election, the inept lieutenant governor, a man named Colden, publicly read to the assembly a letter from Lord Dartmouth in London ordering Colden to prevent the election of any delegates to any and all future congresses. It was an absurd order—unenforceable, arrogant, and to the colonists—even to many Loyalists—infuriating. The letter so incensed New Yorkers that the freeholders vote was 826 to 163, overwhelmingly approving the slate of deputies to choose the delegates to the congress.

On Friday, April 21, the deputies selected the delegates to the congress. King George III had just driven his most loyal supporters into the waiting arms of the patriots, and Captain James DeLancey's power base was gone—as was the only defense of a way of life for a very privileged class of Americans against an increasingly determined repressed majority.

Indeed, when he heard Governor Colden read that absurd order, the captain knew he was beaten. He felt his country was being run by a violent mobocracy that he didn't want to live under. So he sold his magnificent stud, packed his family, and prepared to sail to England. By the time Bissel arrived on April 23, the captain and his family were gone, never to return, leaving Isaac Sears on the docks of New York City, shaking his fist at the empty ocean.

* * *

The captain's uncle Oliver, biding his time on board the HMS *Asia,* laid plans to wage war on the radicals by forming three Loyalist cavalry units. He had no intention of leaving.

On the Friday following Israel Bissel's ride, Sears placed himself at the head of a street gang of 350 men and went in search of Andrew Eliot, a customs collector. The mob seized him, took his keys, and declared the port of New York closed.

That afternoon, the mobs reached a new frenzy when copies of a Philadelphia newspaper arrived at the coffeehouses accusing Loyalist political leader Oliver DeLancey and his cohorts of asking England for troops to deal with the obstreperous patriot mobs. The taverns overflowed with talk of lynching Loyalists.

A few weeks later, the patrician party had its last shred of power cut from under it when the radicals' Provincial Congress usurped the powers of the New York Assembly. The assembly thereafter ceased to meet, thereby ending royal power in the colony. The Provincial Congress was the de facto government until 1777, when the new state constitution was ratified. That same year, Oliver DeLancey's estate, Bloomingdale, would be sacked.

A few weeks after the rioting began in New York, British authority suffered its ultimate indignity.

With mobs smashing open arsenals and streaming aboard British ships to confiscate military stores, the small British garrison of redcoats in Manhattan—loathed by most of the people in New York, even the Tories—was suddenly friendless and in great danger of attack. It had a stormy history of brawling with the Sons of Liberty, mobs on the piers, and, in drunken misconduct, insulting passersby on the streets. The commanding officer decided to move the entire garrison to a British man-of-war in the harbor.

For King George III, there could have been no greater sign of the humbling of British power in America than to have the British Union Jack pulled down from its Manhattan flagpole, followed by the complete withdrawal of that small British garrison.

But a greater humbling was still in the offing. When the soldiers marched from their garrison with five wagon loads of weapons and military equipment to board the man-of-war in the harbor, the triumphant patriots, calling down expletives and curses on their lobster backs, manhandled them and seized the five wagons, as well as the soldiers' baggage, their ammunition, and even their muskets. The helpless British troops made no effort to stop them. Soldiering grimly through the crowds, stripped of all but their uniforms, they were grateful to make it to their ship without any major casualties.

But they didn't forget. In September of the following year, after Howe defeated Washington in Brooklyn and occupied Manhattan, jubilant Loyalists went through the town painting large *R*'s on the front doors of the radicals, inviting reprisals by the vengeful British troops.

Soon after that, Jemmy Rivington set up a new journal and enjoyed heady times firing broadsides at the humorless rebels year after revolutionary year.

After the riots, Alexander McDougall, dropping politics for war, promptly bestirred himself to form a militia. It was a shame in a way. He'd finally finished his apprenticeship in politics.

Isaac Sears was left with nothing to flog but memories. He formed an irregular posse of horsemen, dubbed himself "Colonel" (which outranked "Captain") and galloped up and down Westchester during the Revolution with a band of near criminals to no great effect.

At the same time, the Livingstons were lining up increasingly with the radicals. They led the fight against the Anglican Church. They gave middling support for the popular vote, firm opposition to Parliamentary taxing programs, and solid support of and participation in both the Stamp Tax Congress and the Continental Congress. They saw eventually that both the king and the Loyalists had to go. And they expected to provide the leadership to replace them.

But they didn't realize how much they would have to share that leadership with outré groups and lower classes. They were not at all sure that they wanted to go so far as many others, but circumstance, tolerating no ambivalence, swept them along with all of Captain James DeLancey's other enemies. Only in 1783, at the end of the Revolution, would they see how far they had come, how much they had given up.

The clearest sign of the times occurred during the Revolution when former employees of the Livingstons came to the front door of the mansions and knocked to be admitted. They had come to announce the size of the Livingstons' tax bills, and they then waited with outstretched palms to collect them.

By the turn of the century, almost all of the huge clan of Livingston politicians were out of office. The Livingstons were largely businessmen thereafter.

Philadelphia

From New York City, the post rider ferried across the Hudson River, hastened down the length of New Jersey, setting crowds agog at every settlement and town, and then ferried across the Delaware River at Trenton into Pennsylvania, where he followed the road to Philadelphia—in all a one-hundred-mile ride. It was April 24.

Among the first in Pennsylvania to hear the news was one of that state's leading politicians, Joseph Galloway. He was hiding in his country home north of Philadelphia for fear that the minions of Sam Adams would at any moment come and hang him from his doorstep. They had already mailed him the noose they intended to use.

The forty-four-year-old Galloway had been brooding for weeks. He had almost turned the colonies away from the path of revolution a few months before at the First Continental Congress with his Galloway Plan of Union. And he had been building a coalition for presenting it again at the Second Continental Congress—until he heard about Lexington and Concord.

A lawyer, a very successful businessman, and the political leader of the powerful Quaker coalition in the Pennsylvania legislature, Galloway had been for decades the leading political power in Pennsylvania, a political partner of Ben Franklin, and most recently a dangerous obstacle to Sam Adams.

From a wealthy family, and with the easy manners of a born gentleman, Galloway had all the political assets: abundant funds, poise and surefootedness in society, high intelligence, high motivation with a clear-cut agenda, and a gift for political strategy. A powerful, persuasive speaker and a compelling writer in an age of writers, Galloway had climbed to the very pinnacle of power in Pennsylvania and held it for many years as speaker of the assembly.

Well-to-do in his own right, he had enhanced his financial standing considerably when he married one of the great prizes of Pennsylvania, Grace Growden, daughter of Lawrence Growden, justice of the Pennsylvania Supreme Court and possessor of one of the great fortunes of that colony.

John Adams, feeling that Galloway had an "air of reserve, design and cunning," recognized an adversary when he saw one.

Galloway had gone to the First Congress determined to stop Adams's drive toward revolution by presenting his own plan for a new relationship with Britain. He was aware that it would take all of his political skill to stand up to the brilliant and brutal political juggernaut of Sam Adams. He knew he was going into the biggest battle of his entire life.

It was not a very cohesive congress. Political opinion ranged from angry to indecisive. "One third Tories, another of Whigs, and the rest mongrels," growled John Adams. With "Trimmers & Timeservers" on every side.

A number of delegates, especially those from Captain James DeLancey's New York, had come with the express purpose of stonewalling Sam Adams. Although the moderates and fence-sitters outnumbered the radicals, the question of which faction would seize control of the congress and push through its program had nothing to do with numbers. It had come down to the political will of either Sam Adams or Joseph Galloway.

From the outset Galloway had no illusions about the adversarial skills of Adams. After observing the man at close quarters, he concluded that Sam Adams "eats little, drinks little, sleeps little, thinks much, and is most decisive and indefatigable in his objects." Galloway was well warned. But nothing in Galloway's broad political experience prepared him for what happened next.

Samuel Adams's first step upon arriving at Philadelphia was to visit the docks and piers of the riverfront with a local politician, Charles Thomson, who liked to describe himself as the Sam Adams of Philadelphia. Adams spent some time on the docks and in the taverns talking with the workingmen there, pressing the flesh and preaching his incendiary politics. He quickly won many converts—and lined up some muscle.

Next, without warning, he attacked Galloway directly on the seating of the delegates. As political head of the host city, Galloway had planned to seat the congress in the Pennsylvania State House—later called Independence Hall—which was ideally suited for the congressional sessions. But it was also Galloway's turf.

Despite objections, mainly from the New York delegation, Samuel Adams pressed for meeting in the smaller, somewhat cramped quarters of the more populist, recently built Carpenter's Hall because the workingmen of the city identified with it. Galloway felt that Samuel Adams had slapped his face. He had. Galloway needed to be on his guard.

Samuel Adams's reputation had preceded him, and he shortly had a good estimate of the hostility that was going to be directed toward him. The other delegates had many reasons for distrusting him. Most regarded him as little more than a mob leader who would turn his violence on them as quickly as he had on the merchants and Loyalists in Massachusetts. Others rejected his callused-hands populism. Many delegates, primarily Southerners, distrusted him—and all New Englanders, in fact—for hating Episcopalians. And they all believed that he was secretly bent on independence—"an idea as unpopular in all the middle and south as the Stamp Act itself."

To negate those attitudes even before the Congress officially convened, Samuel Adams quickly formed a coalition with the radical Virginians Patrick Henry and Richard Henry Lee and South Carolina's fervent revolutionary, the wealthy planter Thomas Lynch, and his associate, Gadsden.

Adams quickly saw that the Virginia delegation deferred all questions to Washington. Thereafter, he worked through him. In fact, he held numerous private meetings with Washington to prearrange strategies and votes before the official meetings.

Next, as the first order of business Sam Adams managed to get himself chosen as temporary secretary of the congress. He then got the fiery Lynch to make a nomination for the powerful seat of permanent secretary. Joseph Galloway was almost paralyzed with shock

Joseph Galloway

when he heard the name. It was Galloway's great enemy, the Pennsylvania radical Charles Thomson. Galloway jumped to his feet, objecting strongly. So did John Jay. Duane expostulated, remonstrated, wagged a finger. But the deed was done. Thomson was elected—unanimously, said the record archly.

"Both of these measures, it seems," Galloway wrote regretfully to William Franklin, "were privately settled by an Interest made out of Doors." In short, he'd been jobbed by Adams, lobbying in the taverns. This was the first time many of the conservatives and Southerners had seen Sam Adams up close, and what they were seeing made them shift uncomfortably in their chairs.

With control of the permanent secretary's chair, the radicals next captured the president's chair when Peyton Randolph of Virginia was elected to fill it. Feeling the noose tightening, Galloway noted that it was no coincidence that workingmen from the docks were loitering on the grounds around Carpenter's Hall in an intimidating manner.

Adams would now reach for the dramatic event he had planned before leaving Boston. As he had expected, the congress soon ensnared itself in bickering over ambling verbiage—Shall we call on the Law of Parliament or on the Law of Nature? Matter of right or matter of necessity?

It was dripping hot, the wigs were itchy, many wore wool, stockings were wet with sweat, clouds of flies from the stables intruded, and the congress was drifting. Lawyerly nitpicking droned. Timing was everything, and at that moment something dramatic was needed. Abruptly, seemingly on cue—it was—Paul Revere in great haste galloped into the courtyard of Carpenter's Hall, slapping 300 miles of dust off his clothes as he handed in a document called the Suffolk Resolves.

Before he had left Boston, Sam Adams had prepared for dominating the Continental Congress with this document. He had summoned fifty of the most radical men in the colony, chairmen and members of various committees or of the Sons of Liberty, to a meeting in Coolidge's Tavern in Watertown.

There he arranged for his radical protégé and student Dr. Joseph Warren to write, under Adams's direction, a series of truculent resolutions—including an embargo on British goods—that the fifty men would then push through all the town meetings in Suffolk County, Massachusetts, which included Boston. Those Suffolk Resolves then would be carried to Philadelphia smoking hot, ready for the congress's quick endorsement.

With all the delegates in their seats, silently waiting, President Peyton Randolph began to read the document aloud to their astounded ears. The Suffolk Resolves bristled with accusation—far more so than anything the delegates had ever heard before: "Arbitrary will . . . military executioners . . . murderous law . . . villains . . . our enemies . . ." They shook a fist at the king and called for a new embargo on British products—in effect, closing English ports in retaliation for closing the port of Boston.

Furthermore, the resolves urged every Suffolk community to raise a militia "to learn the arts of war." The "Intolerable Acts" passed by Parliament were "gross infractions," and therefore they deserved "no obedience" from Americans. Judges who took their salaries from the Crown rather than the colonial legislatures held office "unconstitutionally." All the king's taxes were now to be paid to the treasury of an independent provincial government.

In ringing phrases—"the dearbought inheritance of liberty . . .

most sacred obligations . . . unfettered by power . . . our innocent and beloved offspring . . . persevering opposition . . . to merit the approbation of the wise, and the admiration of the brave and free"—the resolves were a rejection of British authority and a defiant call to action.

Peyton Randolph finished reading and raised his head.

Abruptly the room teemed with applause. Radical delegates arose to thump the Massachusetts delegation on their backs, to wring their hands.

Duane, of New York, and Galloway, stung, were on their feet, objecting.

"For Congress to countenance such a statement," Galloway cried, "is tantamount to a complete declaration of war!"

Only the conservatives were listening to him, but, though numerous, they were unorganized; they had been totally outmaneuvered and now they were intimidated by the mob muttering around Carpenter's Hall. They sat in silence.

Congress approved every word, recorded it as a unanimous vote—it wasn't—and, full of self-praise, adjourned for the day.

Galloway had to be simultaneously furious and impressed by Adams's political skill. But there was more to come.

Next, Samuel Adams introduced a proposal that would require the other colonies to come to the aid of Massachusetts if British regulars attacked its citizens. Adams encountered one hitch here: Many delegates warned him to his face that they would come to the aid of Massachusetts only if the British regulars fired first. If his waterfront minions started a shooting confrontation, Massachusetts would go it alone.

Adams now had almost everything all wrapped up just as he had planned. The conservatives had not been organized enough to stonewall him after all. And none of the other delegates seemed to notice that this congress had produced no plan, prepared no counteroffer to the Crown—there was no earnestly pressed program to resolve their differences peacefully, nothing for the other side in London to chew on. Just as Sam Adams had intended, this congress had put nothing on the negotiating table except a loaded musket.

If the delegates really wanted peace and harmony, wanted some new footing with the Crown, which most of them would have said they wanted, then the very last document they should have sent to the king was the Suffolk Resolves. As Galloway had said, it was a declaration of war.

But there was a plan, Joseph Galloway's Plan of Union. Although peppered with flaws, this plan still could have been the basis for serious negotiation. Before it could get a fair hearing, however, Samuel Adams went on the attack.

When he was finally invited to present his plan, Galloway was aware that he had been seriously outmaneuvered by a master politician. But he felt he wasn't beaten. To him, his plan was so right, so obvious, so self-evidently workable, despite Adams's political shenanigans, that it would be irresistible to the delegates.

Galloway was a powerful speaker, and he had a good idea—an American Parliament, the precursor of the dominion constitutions that other British colonies would adopt in the nineteenth century. And it could have worked. Much of Galloway's plan was an outgrowth of the seminal mind of Ben Franklin, who had presented his own Albany Plan of Union back in the 1750s as a way of getting the colonies to act in concert on certain matters. Unfortunately for Franklin's plan, the touchy colonists had not been ready to work together on anything in those early days. But it was now 1774, and the fuse was smoldering. Galloway saw two alternatives: either his plan and peace or Adams's resolves and revolution.

He got up to speak. Logically, rhythmically, persuasively. And he soon had heads nodding in agreement. Here, he said, was the solution to our problems with Britain, a way to resolve the intractable issue of taxation, a way to put colony and king on a new course of harmonious relations. The colonies could avoid a ruinous policy of nonimportation and nonexportation, could get the port of Boston reopened and the Intolerable Acts repealed, and could completely eliminate the danger of civil war.

A number of delegates liked the plan; many were relieved to find what they perceived to be a moderate way out of their dilemma. The Galloway Plan garnered the interested support of James Duane and John Jay. Edward Rutledge of South Carolina declared it to be "almost a perfect plan." Galloway sensed that it was very near to adoption.

Samuel Adams, however, was in no mood for reconciliation or "Union." The port of Boston had been closed by an angry king who had sent an angry army and demanded reparations for the soaked tea. Samuel Adams was damned if he would let anyone pay for it.

There would be no talk of reconciliation until Parliament lifted the Boston blockade and repealed the Intolerable Acts. And maybe not even then.

Patrick Henry and Richard Henry Lee led the attack on the Galloway Plan. They railed against a fatal flaw in the plan as they perceived it. The Parliament in London would have veto power over the acts of the American Parliament, which put everything right where it had been, giving Parliament taxing power over the colonies. Even so, this was a point that could be eliminated or modified, one of the many details that could have been corrected if the delegates had wanted to hammer the plan into a serious proposal. The delegates were leaning more and more toward the Galloway Plan.

Samuel Adams acted—as usual, through indirection behind the scenes. He used his Boston tactics on Galloway. Circulating outside the hall among his new friends from the waterfront, he spread rumors that, inside, Galloway was selling out their liberties. Galloway himself became quickly aware of the muttered threats by dockworkers outside the door, and rather than be tarred and feathered, he drew back. By just one vote, 6 to 5, the Galloway Plan was laid "on the table for future consideration." There it died.

The Congress turned to other matters, hardly aware that it had passed up the last best chance to stop a war.

Later, when Galloway and the supporters of his Plan of Union were absent, Sam Adams passed a motion to remove all records of the Galloway Plan debate. As far as Secretary Charles Thomson's official records were concerned, the entire episode of the Galloway Plan never occurred. This was pure Adams—he'd done it before in Boston. The Galloway Plan would now never be resurrected, and the congress was made to seem to be more in accord than it was. But Sam Adams wasn't finished with Galloway yet. With the plan out of the way, Adams could turn to his prize goal—putting some teeth into the boycott that was called for in the Suffolk Resolves.

For three weeks, the delegates debated over the boycott of English goods. Objection after objection was raised. Whole fishing villages would starve. Businesses by the hundreds would be driven to the wall. Unemployment would be everywhere. The boycott would bring a great depression on the colonies.

Relentlessly, the Liberty men pressed forward, all day in the congress, all evening in the taverns, hammering out, making concessions, smoothing the brows of the troubled, and getting what they wanted.

Finally they had an agreement called the Continental Association. It had fourteen sections detailing the import products that would be banned and the export products that would be exempted from the ban. Tar, pitch, lumber, flaxseed, rice, tobacco, molasses, sugar, coffee, slaves, rice, indigo—each in its turn was examined and listed, yes or no.

It was only after South Carolina walked out of the congress that the radicals agreed to exempt Carolina rice from nonexport rules. And to placate a restive Virginia, the embargo was rescheduled—it would not start until the following year, in September 1775.

This document was no humble request to the colonies. It created a "people police" to enforce the act. Teeth that could bite. The instrument required that "a committee be chosen in every county . . . to watch the wharves and customhouses . . . to the end that all such foes to the rights of British-America may be publicly known, and universally condemned as the enemies of American liberty." One could smell the bubbling tar, see the goose feathers floating in air, watch the arsonist's flaming brand in the dark, firing warehouses and ships. Conform or else.

During the previous boycott, many businessmen, seeing that others were breaking the boycott, smuggling products, and amassing fortunes, had refused to continue. Despite Sam Adams's threats and attacks, the boycott had collapsed. Sam Adams never forgot, never forgave, and never trusted the merchants again. Now Sam Adams had a boycott that could be enforced on every businessman in every town in every colony. Like it or lump it. Obey or burn. This was a boycott that would be enforced by the common man, represented by the dockworkers lurking outside the hall.

Fifty-four men, in illegal assembly, had passed an illegal law that was to be obeyed by every single person of the 2.5 million people in America. The delegates had given the most dangerous man in America the most dangerous weapon he had ever held.

The instrument had a lot of strong—if disorganized—opposition and passed only by a majority vote. A large minority demurred. On October 20, 1774, the delegates assembled to sign the document—many with the greatest reluctance. Under the baleful eye of Sam Adams, Joseph Galloway, his pen trembling with fear, "almost weeping with rage and frustration," scratched his name on the document. Charles Thomson's official records showed the vote as unanimous.

An angry, outmaneuvered Loyalist from Maryland bitterly noted:

"Adams with his crew, and the haughty sultans of the South, juggled the whole conclave of delegates." John Dickinson said that Sam Adams had "forged a network of insurrection" not to be broken by George III, Lord North, and the British military powers.

The Suffolk Resolves calling for embargo were sent to the king with a contradictory resolution asking for reconciliation.

Samuel Adams had gotten everything he had come for: control of the First Continental Congress, defeat of the Galloway Plan and the silencing of Joseph Galloway, passage of the Suffolk Resolves with an enforceable embargo of British goods, and a promise of support from other colonies in case of a shooting confrontation with British troops. He had manipulated the congress past the crossroads of reconciliation and down the path of rebellion. His work done, Sam Adams wanted to hasten back to Boston and the final irrevocable step toward revolution—forcing Gage to attack him. He was now ready for Lexington, six months away.

In a cloud of euphoria, the First Continental Congress disbanded. It was Wednesday, October 26, 1774.

That evening a gala banquet was held in the City Tavern and hosted by the citizens of the city to honor the delegates. Some 500 people drank thirty-two formal toasts, each punctuated by the firing of a cannon on the green. Astonishingly, the last toast expressed the hope that a joyful reconciliation could be reached with the mother country.

Most of the delegates still did not realize that they had taken a major leap toward revolution. They actually believed that the Continental Association agreement—the boycott—would bring Parliament to its knees and force it to repeal the Intolerable Acts, open the port of Boston, and restore things to the way they had been. Richard Henry Lee promised John Adams that Boston's troubles would be over by June—as soon as the mails could crisscross the ocean.

Both Sam Adams and John Adams shook their heads in disbelief. The king wasn't going to fold. He was going to fight. The Continental Association was not a document of persuasion and peace. It was a declaration of war. At the noisy gala, among drinks and toasts and clouds of tobacco smoke, and the hearty laughter of the self-pleased, few—other than the two Adamses and Joseph Galloway—saw this.

Galloway's own words would come back to haunt them: "For

Congress to countenance such a statement is tantamount to a complete declaration of war."

Congress, waving a cageful of doves, had set the colonies on the road to Lexington and Concord.

"I expect no redress," John Adams privately told Patrick Henry. "We must fight."

"By God," answered Henry, "I am of your opinion." The two leading firebrands of America were in accord.

Most of the delegates rested for several days before starting home. It must have been on the various roads out of Philadelphia that on sober reflection many of the delegates finally realized what they had done.

There was one consequence of the congress's action still to be disclosed: "The instrument called the Continental Association, in the name of American liberty, was actually curbing the liberties of the individual."

Summer was gone. So was that terrible Philadelphia heat. In a chilly autumn rain, John Adams and Sam Adams loaded their trunks onto their dripping carriage and left for Boston. As their carriage clattered over the wet paving on its way out of Philadelphia for the ten-day journey home, John Adams thought he was looking his last on Philadelphia. He never expected to return to that city or to congressional business again.

If Sam Adams had had a Rabelasian sense of humor—he didn't—the carriage would have rocked with laughter all the way back to Boston. For he had conned some of the smartest businessmen and leaders in the colonies—most of whom had gone to Philadelphia specifically to stonewall him—into doing the last thing they had wanted to do. Instead, jouncing next to his cousin, Sam Adams left Philadelphia grimly satisfied that he had finally finished off Joseph Galloway and all thoughts of redress and reconciliation.

True, Galloway's hopes were fading. But with Adams gone, Galloway felt he could return to the battle. He still had several important moves left. And he thought Sam Adams wouldn't notice.

All through the bitter winter that followed, Galloway stared into the flames of his fireplace and smoldered over the tactics of Sam Adams—as had hundreds of Adams's victims before him. Defeated in the congress, he would appeal to the public; turning to his pen, he discovered a gift for political polemics.

Galloway started by taking a piece of paper and a pen and writing "A Candid Examination of the Mutual Claims of Great Britain and the Colonies: With a Plan of Accommodation on Constitutional Principles." He spent much of the winter drafting the pamphlet, and when he finished he had become one of the most important writers in the war of ideas.

First he sounded a solemn warning. He knew that many colonists believed that Sam Adams secretly wanted independence and that he wanted it so Boston could take over all thirteen colonies. Galloway played to that fear. The country, he warned, was in danger of having armed mobs rule.

The preamble then mounted one of the most fervid attacks on Sam Adams and his political methods ever written: ". . . freedom of speech suppressed, the liberty and secrecy of the press destroyed, the voice of truth silenced, a lawless power established throughout the colonies . . . depriving men of their natural rights and inflicting penalties more severe than death itself."

He warned his readers of the dangers of letting the country be run by unlettered mobs led by demagogues. Then he laid before his readers as a means of salvation his Plan of Union, essentially as he had presented it to the congress.

Galloway went up to New York City and got James Rivington to publish his pamphlet. He then actively circulated the plan, sending copies to people of influence, especially those who distrusted Sam Adams and most especially those in New York City, where the strongest resistance to Adams existed. In fact, he spent some weeks during the winter in New York City personally cultivating more support—and spreading the word against Sam Adams.

His pamphlet had a powerful effect on many readers. It offered a way of moderation in the quarrel with England and avoided the violent sundering that the radicals were pressing for. Few people, very few, wanted the economic agony of an embargo. Fewer really wanted independence. And still fewer wanted war. Most wanted urgently to avoid the violence and bloodshed war would entail. John Adams himself estimated that no more than a third of the people wanted independence. Others felt he estimated on the high side. Ten percent—if that—was said to be a more probable number.

Still ready to fight, Galloway returned to the Pennsylvania Assembly where he began to reassert some of his old powers. Summoning his greatest oratorical skills, he came within a whisker of getting the

Pennsylvania Assembly to vote a petition of reconciliation to send to the king—just as Captain James DeLancey was doing in the New York Assembly.

It was only a matter of time before Sam Adams heard about Galloway's activities. In the mail, Galloway found a gallows noose. A day later, he received an insurance policy indicating that he would not live six days longer: "Hang yourself or we will do it for you."

"Every night," Galloway later told a committee of the House of Commons, "I expected would be my last. Men were excited by persons from the northward [read Boston], by falsehoods fabricated for the purpose, to put me to death. Several attempts were made. I declared my innocence in the public papers in vain."

He felt so intimidated that he fled from his Philadelphia mansion to his country estate in the village of Trevose, Bucks County, where, he told the parliamentary committee, he lived in "the utmost danger from mobs raised by Mr. Adams of being hung at [my] own door."

He waited all winter into the spring for his pamphlet to take effect.

But before the pamphlet could work its way into the minds and hearts of the colonists, before the full reaction of thinking men could be registered, the Galloway Plan was brought down by the sharpest and most convincing argument against it: the fighting at Lexington and Concord. England was now cast in the light not of a mother to be reasoned with but of a tyrant bent on bloodshed and brutal suppression. Any thought of union was past.

For Galloway, even more terrifying was the coming of May and the opening of the Second Continental Congress, which would deal with Lexington and Concord. Most terrifying of all, Sam Adams would be returning to Philadelphia.

Galloway had just one move left. He waited for his old political partner to return home. Benjamin Franklin, after eighteen years in London, was at that moment on the high seas, and would debark just days before the Second Continental Congress was to convene. Together they would have to hatch a strategy to handle Sam Adams before it was too late.

In the meantime, Galloway cowered in his country home, listening for the sound of a lynch mob coming up his walk.

The news did not stop for the likes of Joseph Galloway, however. Sometime after noon, on Monday, April 24, just five days after Lexington and Concord, the express rider entered Philadelphia. He pelted down Broad Street on his lathered horse and into the center of the city, crying out his news as he rode directly to City Tavern, on Second Street at Walnut.

City Tavern, less than two years old, was the most popular dining and meeting place of the city's leaders. It was where the two Adams cousins had done their most serious after-hours politicking during the First Continental Congress. It was also the main headquarters of the whole radical movement. In an upstairs meeting room, the all-powerful Philadelphia Committee of Observation, Inspection and Correspondence was just sitting down when the express rider came shouting up the street.

Hearing his excited voice, they hurried en masse down the stairs, following people from the sumptuous dining rooms who were still dabbing their mouths with napkins and others from the elegant coffee room who were still holding their newspapers. Pushing out onto the street, they all stood astonished at the rider's words. More men came running up from the piers and docks, only a few blocks away. Church bells began to peal. Someone found a drum. From all over the city militia members ran to muster and to parade. Marching flags were unfurled. People hurried through the streets to carry the news to others. Residents in large numbers came out of their houses to confer with each other, intent on extracting the tiniest morsel of significance from it all. Knots of people stood expostulating in the markets and on the docks. The hastily gathering militia units formed an impromptu parade. More drums and fifes sounded, more flags were shaken out.

But at the fringes of the celebration, many others—a large number, in fact—stood in solemn silence. There on the brick sidewalk, the chairman of the Committee of Correspondence, John Dickinson, was clearly one of the most disturbed. To him this news could not have been worse.

A wealthy Quaker lawyer and one of the largest landowners in Pennsylvania, he was, more significantly, one of the most famous men in the colonies as the author of "Letters from a Farmer in Pennsylvania to the Inhabitants of the British Colonies," possibly "the most brilliant event in the literary history of the Revolution." Writting in response to Parliament's Townshend duties, Dickinson, in a prose that was calm and orderly, reasonable and conciliatory, marshaled a compelling parade of arguments that demonstrated that the Townshend duties were clearly unconstitutional under British law. He urged Americans to oppose them in the most firm but most peaceable manner and to impel Parliament to rescind them. The cause of liberty, he wrote, "is the cause of too much dignity, to be sullied by turbulence and tumult." He exclaimed against "any thought of Independence as a fatal calamity." Dickinson called for redress, not separation.

This was the constitutional argument everyone had been looking for, and overnight it made "The Farmer" a hero: "He was saluted by town meetings and grand juries, toasted at patriot dinners throughout the colonies." Entering a courtroom in a legal matter, Dickinson stood quietly as the court's business was interrupted while he received a standing ovation. Songs and poems were written in his honor. John Adams paid Dickinson the ultimate compliment: "He is a true Bostonian."

Published in twelve weekly installments in Philadelphia, "Letters from a Farmer in Pennsylvania" was eagerly greeted and copied by other newspapers. The individual letters were often read aloud in taverns and homes throughout the colonies. Within weeks they were published in pamphlet form that soon went through eight American editions, several in London, another in Dublin, another in Amsterdam and Paris. Voltaire applauded them. In Europe, Dickinson was compared to Cicero. A common toast in the colonies became "To the Farmer."

But that had been ten years earlier. Since then the mood of the country had shifted to such a degree that Dickinson was now almost a conservative. Yet neither Dickinson nor his fellow delegates to the Second Congress had any illusions about their personal safety if their efforts should fail. In Dickinson's ears sounded his distraught mother's terrifying words: "Johnny, you will be hanged, your estate will be forfeited and confiscated, you will leave your excellent wife a widow and your charming children orphans, beggars and infamous."

Dickinson had replaced his political enemy, the fugitive Joseph Galloway, as a Pennsylvania delegate to the Second Continental Congress. He disliked not only Galloway but also his Plan of Union. More than that, however, he deplored Sam Adams's boycott and was determined to see it repealed. Greatly desiring reconciliation with England, he felt that the battle at Lexington and Concord made a discourse with the Crown more imperative than ever.

He planned to present and defend another petition for reconciliation, knowing he was going to get bellowing tiptoe opposition from both the Virginia delegation and the Boston delegation, particularly the paired radicals Sam Adams and John Adams. Yet he felt he could handle Sam Adams's juggernaut, and he fully intended to be more successful than Galloway had been.

All delegates foresaw that a major battle was shaping up between the conciliatory moderates and the war-hungry radicals in the Second Congress.

Charles Thomson was one of the patrons who had come hurrying out of the City Tavern when Bissel arrived. Busy as he was with preparations for the Second Continental Congress, Thomson took the time to bask in the drum-thumping celebrations that had sprung up. The man who styled himself the Sam Adams of Philadelphia was as close to a firebrand as Philadelphia offered. And now he was feeling exultant over the total defeat of Joseph Galloway and his shattered Quaker coalition.

The forty-five-year-old Thomson, a Scotch Irishman and an orphan, who had become a schoolteacher, a successful businessman, then for many years a politician, was well-known for his passionate speeches. He was also quick and smart—but a little too unprincipled for many of his associates, who were put off by his slick political moves and questionable debating tricks. During one well-remembered encounter, Thomson faced an opponent whose shouting was carrying the day. To end the debate and avoid a losing vote, Thomson pretended to faint.

A populist politician, Thomson was a favorite with the workers along the waterfront whom he was trying to mold into a political force similar to the waterfront gangs in Boston. In fact, in his first visit to Philadelphia, during the First Congress, Samuel Adams had given Thomson instructive lessons on just this type of politics. But Thomson, for all his skills, was no Adams, and conservative Philadelphia was no Boston.

Although Philadelphia was subjected to some mobbing, the attempted transplant of radical tactics was never quite successful. As was the case in New York, a large proportion of Philadelphia's population was Loyalist. And unlike the Boston Loyalists, these Loyalists did not tolerate radical intimidation. Quickly outspoken, they loudly challenged radicals in drawing rooms, taverns, and coffee houses—even in the colonial assembly. During the Stamp Act riots in Boston, Philadelphia had remained completely calm because over 800 citizens turned out to patrol the streets and repel would-be rioters. The Loyalists of Boston were never able to act in such concert.

Strongly conservative religious elements in the city—including

Charles Thomson

the dominant Quakers, the Anglicans, and many splinter groups such as the Mennonites—looked with solemn doubt on Thomson's extreme radicalism and had refused to elect him a delegate to the First Congress. So the chastened Thomson consoled himself by marrying a fortune in the person of John Dickinson's cousin, a lady with 5,000 pounds sterling, and went off on his honeymoon.

But Sam Adams had an important job in mind for his eager disciple, and as soon as he arrived in Philadelphia for the First Continental Congress, Adams sent for him. Late on the night before the first day's session, Charles Thomson walked into the bar of the City Tavern and for the first time met his role model and John Adams.

At first glance, they found him to be a "striking looking man, about 45, thin, with piercing, deepset eyes, wearing his own hair cut short above the ears." After a close and intimate tête-à-tête, the Adamses liked his radicalism and found him to be "the very life of the cause of liberty in Pennsylvania." Much to the dismay of Galloway and the voters of Pennsylvania, Sam Adams would soon manipulate

Thomson's election as permanent secretary of the First Continental Congress.

Robert Morris also stood on the brick sidewalk outside City Tavern, unaware that fame awaited him.

A native of Liverpool, at age thirteen he joined his father, a tobacco exporter, in Oxford, Maryland. By 1754, at age twenty, his brilliance in business had earned him a full partnership in a very successful Philadelphia shipping firm. Before he had reached his majority, he was on his way to a great fortune.

Now forty-one years old, he was one of the busiest men in Philadelphia. Because of his great administrative skills and his reputation as a financial wizard he had been selected to be a delegate from Pennsylvania to the Second Continental Congress, where his special talents would be used on the committee for financial affairs and military procurement. He was also serving on the Pennsylvania Council of Safety, the Committee of Correspondence, the Provincial Assembly, and the state legislature, and soon he would be even busier helping his firm fulfill a contract from the Continental Congress to purchase arms and ammunition.

He would be one of only two men to sign all three founding documents of the country—the Declaration of Independence, the Articles of Confederation, and the Constitution (Roger Sherman of Connecticut was the other). But his greatest service to his new nation was as the man whose financial legerdemain would save the Continental army from disbanding, and thereby would save the colonies from ignominious surrender. For this he would later become known as the financier of the Revolution.

As the crowd around the express rider grew, more church bells began to peal. Even before the news arrived, Philadelphia had been agog with talk and arguments. It had been preparing itself for the arrival of more than fifty delegates to the Second Congress, many already in transit with their supporting groups. With the gun smoke of Lexington and Concord wafting over their deliberations, their decisions in the next few weeks would be critical.

Now, suddenly, in every tavern and coffee shop and home everyone had an opinion. Everyone had a plan. Everyone disagreed with everyone else. Lexington farmers shot dead by British soldiers. It was almost too much to take in.

* * *

Robert Morris

Philadelphia was the ideal setting for the Continental Congress. Centrally located among the colonies, it was a comfortable, gracious city, welcoming the road-weary or sea-weary traveler with many inns and taverns and the best of colonial foods, drinks, comfortable beds, convivial company, theaters, bowling on the green, libraries, music, dancing, and other entertainment.

John Adams who had spent much time in the City Tavern during the last congressional session, called it "the most genteel tavern in America. Spacious, airy, with expensive furnishings in the latest London fashion, several large meeting rooms and a ball room, as well as its bar and two kitchens, a handsome coffee room stocked with English and American papers and magazines."

The very popular Indian Queen, on Fourth Street, run by Quaker John Biddle, offered excellent food and good liquors, which were not sold after eleven P.M. It attracted patrons such as Thomas Jefferson. Washington and Hancock, both of whom accepted only the best, preferred the old Slate Roof House, run by Widow Graydon, while the gourmets flocked to "the magnificent Oeller's Hotel for dinners at sixth and Chestnut opposite Congress Hall famous for its punches and its sixty-foot-square Assembly Room." Always popular and full of talk were the Library Tavern and the Centre House, which claimed the best bowling in town.

Businessmen often preferred the London Coffee House, at the southwest corner of Market and Front, one of the largest and most popular coffee houses in the city. Every day at noon businessmen gathered to hear the latest ship reports, commodity prices, and news. It had English and American newspapers, and on Post Days it received reports from special correspondents writing from all over, including especially London. Its meeting rooms on the second floor were always in demand.

The innkeepers and taverns looked forward to the till-filling visits of all the delegates except for one man. John Hancock, who lived for pomp and circumstance, was coming for the first time to Philadelphia. Clattering noisily through the streets of the city in his gilded carriage, he would ride preceded, flanked, and followed by a cadre of mounted horsemen. Restaurateurs would hate to see them coming for they were raucous, unsettling, quarrelsome, disturbing, and given to consuming great quantities of food and drink—especially drink. Then they would gallop off into the night without paying the bill.

Other attractions for the delegates were the homes of the wealthy, homes eager to entertain these famous men from "foreign" colonies. John Adams described a meal at the home of Mr. Miers Fischer: "ducks, hams, chickens, beef, pig, tarts, creams, custards, jellies, fools, trifles, floating islands, beer, porter, punch, wine—and a long etc." At the home of Philadelphia's wealthy Mayor Powel, Adams was served "curds and creams, jellies, sweetmeats, twenty sorts of tarts, fools, trifles, floating island, whipped sillabubs, Parmesan cheese, punch, wine, porter, beer . . ."

This ability to so easily accommodate the delegates resulted from Philadelphia's being the largest, the richest and the most sophisticated city in the colonies. It was among the largest cities in the entire British empire—second, in fact, only to London in population. It had nearly 40,000 residents, 6,000 houses, and 300 shops, compared to 16,000 people in Boston and 25,000 in New York. It was known for its high culture, colleges, medical establishment, libraries, and learned societies—including the Philosophical Society, which exchanged papers with similar societies in London and Europe. Philadelphia could even boast a Society of Astronomers headed by an internationally famous astronomer, David Rittenhouse. It was also the home of one of the world's leading botanists, John Bartram, royal botanist to the king; home also to one of the leading scientists in the world, Benjamin Franklin, and home, too, of the multifaceted painting family of Charles Willson Peale.

A major banking center, a leading seaport crowded with ships and recording shipbuilding volume far beyond Boston's, Philadelphia was also the capital of Pennsylvania, and visitors came to stare at its state house (later renamed Independence Hall), the meeting place of the Pennsylvania Assembly and the largest building in the colonies. The city also boasted several other features that particularly impressed visitors: paved streets, bricked sidewalks and streetlights. Its influence among the other colonies was enormous.

Conservative and firmly antiwar—this Quaker stronghold was a city of churches crowded on Sundays. It also led all other cities in smuggling.

The Quakers, who founded Pennsylvania and still dominated it, represented only a third of the population but did nearly three quarters of the trade. Parsimonious, shunning ostentation, yet avid as businessmen, they were firmly conservative and nonviolent.

Quaker belief opposed any disobedience to the king and did not permit armed conflict, especially armed conflict against constituted authority. Hence the Quaker meetings were absolutely opposed to any revolutionary movement. This position would cause increasing social strains with their more violent non-Quaker neighbors, who were about to fight bloody battles all through the Revolution—two of them right on Philadelphia's doorstep, the battle of the Brandywine and the battle of Germantown.

Behind closed doors, the Quakers, who called themselves the Religious Society of Friends, were about to undergo the most tumultuous soul-searching and quarreling in their history. It was going to take the most determined effort to emerge from the American Revolution in one cohesive piece.

The divisions that were growing among the Quakers were apparent even in the number of horse-drawn carriages in the city. In 1772 there were eighty-four, which was surprisingly few for a city of Philadelphia's size and sophistication; New York, which was much smaller, had sixty-nine. Most Quakers, even the wealthy who could easily afford them, frowned on carriages as marks of great vanity and pomp and refused to buy them. But thirty Quakers did own carriages, albeit the least pretentious and therefore the most practical kind, the unadorned coach wagon or buggy.

To Samuel Wetherill, a Quaker textile manufacturer, the news meant war, bloody, ruinous war, against the strongest monarch since the greatest days of the Roman Empire. It meant invading armies, cities laid waste—a vengeful king meting out punishments.

Friend Wetherill realized in an instant that Lexington put the Religious Society of Friends—the Quakers—in a dilemma that could tear the Society into pieces. The moral and ethical implications were staggering: Should they—*could* they—remain neutral as they were counseled by the Peace Testimony issued by the yearly meeting?

Their Divine Experiment in Pennsylvania had been in turmoil for decades, constantly challenged by events. In 1756, the Quakers had objected so vehemently to Pennsylvania's participation in the French and Indian War that they withdrew from representation in the Pennsylvania Assembly. Literally, they had shunned their own government. What should they do in a war against their own king?

Wetherill, a recorded minister of the Fourth Street Monthly Meeting, came from a long line of dissenting Friends who had quarreled many times with their own meetings, going back to the early seventeenth century. He was also a manager of the United Company of Pennsylvania Manufacturers. As a Quaker and a businessman, he knew he was caught in the middle. Armies need uniforms; uniforms need cloth. If asked to use his weaving talents for the cause, what would he do? Order his textile factory to make cloth for the uniforms for the Continental army? If he did he knew it meant being disowned by the pacifist Society. He would thereafter be shunned by his lifelong companions, shut out from a whole way of life. If he didn't, his non-Quaker friends and associates could accuse him of betraying the American cause. They could even throw him into prison.

Wetherill and other Quaker businessmen had already risked disownment from their meetings. Going against the yearly meeting's council of strict neutrality, they had actively condemned the Stamp Act, opposed the Townshend duties, and fully participated in the first nonimportation program against British goods, which had failed. But that had been politics. This was far more significant. This was war.

Many Quakers had already made their choice between their faith and their country.

Thomas Fisher, a dedicated young Quaker, was pushed into a particularly personal dilemma by the news from Lexington and Concord. The divisive issue that tormented Quaker affairs during the Revolution centered on a single overwhelming question: Was it possible for Quakers to continue allegiance to the Commonwealth against England without violating their own pacifist principles? Fisher knew he would not yield on his pacifism and wondered how he could continue to live within his Commonwealth that was warring against the mother country. Soon he would get his answer: Self-styled patriots would send him and a band of other prominent Philadelphia Quakers into exile in Virginia just before the British occupation of that city. He would not be allowed to return home until April 1778.

Timothy Matlack, owner of a hardware business, had been disowned by his Quaker meeting. His transgressions were his failure to pay his just debts and consorting with "the wrong kind of company." This was caused by his habits, most unusual for an abstemious Quaker: He was given to gambling, horse racing and déclassé cockfighting. His radical political views, which he was inclined to assert in a loud public voice, often ran athwart the views of the more conservative, wealthier Friends. On one occasion, he and fellow Friend Whitehead Humphreys, one of Philadelphia's wealthiest Quakers, were seen tussling in the dust. By all accounts, Matlack won. Elected secretary of the Supreme Executive Council of Pennsylvania, Matlack for a while served on the executive committee of the new state government of Pennsylvania in 1776.

Christopher Marshall, newly retired as an apothecary, was violating Quaker nonviolent tenets by serving as a member of the revolutionary Council of Public Safety, although he had already been disowned by his meeting, which accused him of having counterfeited currency. He might have been guilty. A man of political influence, Marshall had retired wealthy, and few believed he'd made his fortune from his apothecary business.

And drilling with muskets this day in the factory yard were the Quaker Blues, a band of armed Quakers, every one of whom had been read out of their meetings.

It took only a few moments for the news of Lexington and Concord to reach 239 Arch Street, the business place and home of the twenty-two-year-old Quaker seamstress and flagmaker Elizabeth Ross, who was nursing her husband, John, an upholsterer and a volunteer patriot watchman. He had been badly injured the previous summer when the munitions warehouse he was guarding down by the Delaware River exploded.

Born Elizabeth Griscom, she had been disowned by the Quaker yearly meeting because she had eloped, an immoral act in itself, with Ross and committed the unpardonable sin of "marrying a person of another religious persuasion." John Ross was particularly unacceptable since he came from a family of active patriots, two of whom, both uncles to Ross, were to sign the Declaration of Independence—George Ross and George Reed. John Ross was not to live to see his uncles sign the Declaration. By year's end he would be dead of his injuries.

Controversy would continue after Elizabeth Ross's death until today, with historians in fruitless dispute trying to determine whether, as flagmaker "Betsy Ross," she designed the Stars and Stripes for General George Washington or simply sewed the prototype.

The Pennsylvania Dutch—a term applied to German Lutherans and many offspring religions, including Mennonites, Amish, Dunkards, and Schwenkfelders—comprised another third of the city's population, mostly German speaking, insular, and nonpolitical. They showed little inclination to take up arms.

The last third were the Scotch Presbyterians—poor, restive, leading hardscrabble lives in the wilderness counties at the western end of the colony, demanding more freedom and more representation in the state legislature, and always ready for a fight. They provided the bulk of recruits for the Pennsylvania militia.

Despite deep distrust and strong opposition, there was also a growing number of Catholics—indeed, they now had two churches, Old Saint Joseph's Church, in Willings Alley, which once had been the only church permitting Catholic worship in all the colonies, and Old Saint Mary's, at Third and Pine, built in 1763 to handle the increasing Catholic population.

Catholics, regarded with great distrust as instruments of the Pope in Rome, had been largely barred from public office. But during the Revolution, dedicated service to the patriot cause would be given by many Catholics such as Charles Carroll of Maryland. He risked his life and one of the greatest fortunes in America when he signed the Declaration and then went on to work indefatigably for American independence.

* * *

Haym Salomon, a bill discounter and broker, was sitting in the Coffee House, at Front between Arch and Market, which he used for his office, as did other dealers, when the news of the fighting in Massachusetts reached Philadelphia. They all came out on the street to stare about, argue over the news, and wonder what would happen next and what they should do. Salomon would be remembered by patriots for his vital role as one of the financiers of the Revolution.

He was a member of Philadelphia's small Jewish congregation—barely more than 100—who met in the Mikveh Israel Synagogue, at Third and Cherry Streets. No one was more exercised by the ominous news from Lexington. Even as early as 1768, in reaction to the Stamp Act, the few Jewish merchants, nine in all, had put their lives and fortunes in jeopardy by signing the nonimportation agreement along with a number of Christian merchants. Had it failed, the signers could have been tried for treason and hung.

David Franks, a prominent fifty-five-year-old Jewish businessman, could not foresee from the day's news that he would sail into stormy political trouble during the Revolution. A man of considerable wealth, he owned a town house near City Tavern, a store near Christ Church, a country seat near Woodford, four miles out of the city, and a large horse-drawn carriage to shuttle him back and forth on his errands.

He was from a large Jewish merchant family doing business in Britain, Canada, and the colonies. Although he had been born in New York, he lived in Philadelphia with his Christian wife. And he had powerful connections with the Loyalists in New York through his sister, Phila, who was married to Loyalist Oliver DeLancey. He was also related to the powerful Franks family of Montreal. His cousin there, David Solebury Franks, became a major in the Continental army and served as an aide-de-camp to Benedict Arnold. Another Canadian cousin also served in the Continental army.

David Franks himself was later appointed by Congress as a commissary for British prisoners, supplying food and clothing to captured Britons at British expense. Suspected of British loyalties, he was arrested twice—in 1778 and 1780—and finally sailed for England in 1782. He returned to Philadelphia sometime before 1792, and died there in 1793 during the yellow fever epidemic.

American Jews had strong feelings about freedom—and equally strong feelings about anti-Semitic laws and practices in Great Britain. When the British army occupied New York and Charleston and Savannah, rather than collaborate with British army rule, many Jews from those cities abandoned much of what they owned and moved to Philadelphia, giving the Quaker city, for a time, the largest Jewish population in the colonies.

Most Jews were revolutionaries, but not all. Fifteen young men from Jewish families in New York City enlisted in Oliver DeLancey's Loyalist battalion.

When Dr. Benjamin Rush, already one of the most eminent physicians in the colonies at the age of twenty-nine, heard the news from Lexington and Concord, he immediately decided to become a revolutionary. "I now resolved to bear my share of the duties and burdens of the approaching Revolution," he announced. "I considered the separation of the colonies from Great Britain as inevitable. The first gun that was fired at an American cut the cord that had tied the two countries together. It was the signal for the commencement of our independence. . . ."

Enthusiastic, passionate, possessing boundless energy, he posted a string of outstanding accomplishments that would have made the reputations of five or ten men—as a doctor, a medical educator, the first professor of chemistry in America, author of the first American textbook on chemistry, a temperance advocate, a political writer, a reformer-moralist, co-founder of the Pennsylvania Society for Promoting the Abolition of Slavery, co-founder and vice president of the Philadelphia Bible Society, one of the founders of Pennsylvania Hospital (along with Ben Franklin) and of the University of Pennsylvania and Dickinson College in Carlisle, one of the founders of the Philadelphia College of Physicians, founder of the Philadelphia Dispensary, the country's first free medical clinic, a pioneer in the treatment of the insane and author of one of the first texts on treating mental illness, an advocate of prison reform, a campaigner against capital punishment, an active member of the American Philosophical Society, a delegate to the Second Continental Congress, one of the youngest signers of the Declaration of Independence and a champion of the Constitution, and a military surgeon with the Pennyslvania militia—to mention a few.

He started his remarkable career by graduating from the College of New Jersey (Princeton) and from the school of medicine at the University of Edinburgh. He married, and for the rest of his life he cherished his wife, Julia, the daughter of another signer of the

Declaration, Richard Stockton of New Jersey. Dr. Rush fathered thirteen children.

With his endlessly busy professional life, he somehow found time to meet and charm most of the eminent Americans who visited Philadelphia—and in the turbulent 1770s, most did visit at one point or another.

Dr. Rush, with the din of revolutionary celebration ringing in his ears, was about to involve himself in one of his most historically significant acts—coaxing into being probably the single most stunning and influential tracts of the entire American Revolution.

With the streets rattling with drums, someone ran into the offices of the bookseller Robert Aitkin, publisher of a new monthly publication, the *Pennsylvania Magazine or American Monthly Museum,* and proclaimed the news of Lexington and Concord. One of Aitkin's employees, a recent immigrant from London and a failure at everything he had ever touched, pricked up his ears.

Hardly handsome, with a large pockmarked nose, skin stretched over prominent cheekbones, and piercing blue eyes, Thomas Paine, the son of a staymaker in Thetford, East Anglia, like so many others from that period, was taken from school at the age of thirteen and set to learn a trade. For him, it was his father's craft of staymaking. His career from that point was dreary, unproductive, aimless. During the Seven Years' War, he escaped the drudgery of staymaking by serving on a British privateer. By war's end, he was glad to abandon the cutlass. With the peace, reluctantly he returned to making corsets.

At age twenty-two, he married. His wife quickly became pregnant, and at age twenty-three he carried both his wife and his newborn daughter from the birth bed to the burial ground. He became in time a tax man, was accused of goldbricking, and was fired—he had written fictitious reports at home without visiting the sites he was supposed to examine. He worked a short time as an usher in an academy, tried preaching, managed to wiggle back into his tax job, married again, and on the side ran a grocery and tobacconist shop, neglecting his wife, who in turn neglected her marriage vows.

Fired again from the tax job, he watched his possessions auctioned for debt, parted without much rancor from his wife, and wandered into London with nothing exceptional to offer any employer and having only one passionate interest. Talk. Tavern talk. Endless tavern talk.

In London, his luck changed. He met Ben Franklin, who was impressed enough with Paine to pack him off to Philadelphia with a letter of endorsement that recommended him as "an ingenious, worthy, young man." He arrived in November 1774, just days after the First Continental Congress had disbanded and gone home.

A turbulent man, and born a Quaker, Paine promptly responded to the supercharged political atmosphere he found by becoming enchanted with America. At first he had only an outsider's concern with the brewing violence against Britain. But, standing amid the type boxes and printing presses of Aitkin's magazine, and wondering where Lexington and Concord were, he was so morally outraged by the report from there that his loyalty to Britain was shattered on the spot. Instantly.

"No man," he remembered, "was a warmer wisher for a reconciliation than myself before the fatal nineteenth of April, 1775, but the moment the event of that day was known, I rejected the hardened, sullen-tempered Pharaoh of England for ever; and disdained the wretch that with the pretended title of Father of His People can unfeelingly hear of their slaughter and composedly sleep with their blood upon his soul."

Paine was charming, loquacious, fascinating when he talked. In the course of working on the magazine he found himself in amicable relations with some of the brightest and most powerful men in Philadelphia. Invited to the home of Dr. Rush, the fervent new radical held long talks with the doctor on the need to challenge the set ideas that most Americans had about parting from England. What was needed, they agreed, was a pamphlet that would persuade Americans to accept the idea of revolution, a complete break from England.

With added encouragement from Ben Franklin, Paine set out to write and finish the pamphlet by New Year's Day of 1776 as a surprise for Franklin. He didn't quite make it. Several printers felt the piece so inflammatory they dared not set it in type. Dr. Rush finally had to intercede by helping to find a publisher—printer Robert Bell of Third Street in Philadelphia—a "thoughtless and fearless Whig." Rush is even credited with suggesting the title: "Common Sense."

"Common Sense" was published January 10, 1776, nine months after Lexington and Concord and Paine's conversion and seven months before the Declaration of Independence. Its timing was flawless. Public opinion was drifting. There was no philosophical focus. Most Americans were still firmly loyal to the mother country. To others the idea of freedom was appealing but frightening. Few believed it was possible. And many voices warned of the dangers of severing the umbilical cord. Despite Lexington, many still talked of reconciliation. To go beyond Lexington—to present the ultimate defiance to the Crown—required an amount of courage many Americans had not yet found. All understood that they could be hung for treason. Fear must have been a constant presence for many.

The colonies needed a self-assured, defiant voice that could raise their spirits, pull them together, make them believe in themselves. The colonies needed "Common Sense."

"I offer nothing more than simple facts, plain arguments and common sense . . ." So argued Thomas Paine with the plainest language as he proceeded to destroy the idea of kingship over America and to destroy the fear of freedom. He presented thoughts the colonists had never let themselves entertain.

One by one, every idea in defense of the king, of England's rule, of the mother-child relationship that he had ever come across in all the books he had read, in all the taverns he had ever sat in, were plucked up, held on view, scornfully examined, reduced to nonsense, and flung aside. When he got through, common sense showed that there was only one choice left—freedom.

First he ridiculed the accepted idea among colonists that the wise and tormented king loved his American subjects while the Parliament of corrupt politicians wanted to enslave them. The villain in the piece is the king himself, Paine asserted. He then went on to condemn the king, the Crown, and hereditary rule as nothing less than scandalous usurpations of public power by a contemptible human being. The kingship was a tyranny—an absurdity. The peerage was a tyranny. One man bending his knee to another was a wickedness. King George III, "the Royal Brute of Great Britain," controlled England by corrupting the entire constitution and by controlling the House of Commons with bribes and intimidation. British constitutional restraints were a mockery. "The will of the king is as much the law of the land in Britain as in France . . ."

He scorned royalty as "the most prosperous invention the devil ever set on foot for the promotion of idolatry. No man should worship another—yet all of England prostrates itself before the throne and King George III.'' And who was George III? ''A worm, who in the midst of his splendor is crumbling into dust! . . . The state of a king shuts him off from the world, yet the business of a king requires him to know it thoroughly.'' This combination made the king's ''whole character absurd and useless. . . . The king hath little more to do than make war and give away places.'' In fact, the king was nothing more than ''a giver of places and pensions.''

''One of the strongest *natural* proofs of the folly of hereditary right in kings,'' Paine said, ''is, that nature disapproves it, otherwise she would not so frequently turn it into ridicule by giving mankind an *ass* for a *lion*.'' George III was descended from ''the principle ruffian of some restless gang.'' He was ''chief among plunderers. . . . The evil of hereditary succession . . . opens the door to the *foolish*, the *wicked*, the *improper*.''

Calling the king an ass, a principal ruffian, chief among plunderers, foolish, wicked, and improper was not just shocking. It was far more than treason. It was breathtaking. Thomas Paine verbally pulled George III off the dais, flung him to the ground, then kicked over the throne itself. But that was just the beginning.

It took only a few paragraphs to trample into the dust the whole idea of hereditary succession—a major virtue of kingship that was believed to prevent civil wars: ''The whole history of England disowns the fact. Thirty kings and two minors have reigned in that distracted kingdom since the conquest in which time there have been (including the Revolution [of 1688]) no less than eight civil wars and nineteen rebellions . . . monarchy and succession have laid . . . the world in blood and ashes. . . . Another evil which attends hereditary succession is that the throne is subject to be possessed by a minor at any age,'' while just as dangerous was an old king, who, sick ''with age and infirmity enters the last stage of human weakness.''

Paine then attacked the idea of reconciliation and dismissed it as hopeless in the face of an intransigent, stubborn king: ''Nothing flatters vanity, or confirms obstinacy in Kings more than repeated petitioning.'' And who was calling for reconciliation? ''Interested men, who are not to be trusted; weak men, who *cannot* see; prejudiced men, who *will not* see; and a certain set of moderate men who think better of the European world than it deserves; and this last class, by an ill-judged deliberation, will be the cause of more calamities to this continent, than all the other three.''

He dismissed the idea of dependency on England, the need for its protection, the idea of America's military weakness. He condemned the domination of mercantilism, of protective mother love. Britain wanted to keep the colonies not for love and affection but for the most selfish of motives—for the profit of the British economy. He dismissed with contempt the fear of the fainthearted that the colonies were too diverse, too quarrelsome to unite. He even laid out his idea of a constitutional self-government with a president and an assembly.

Separation, he assured his readers—independence—was inevitable and would bring benefits and joys beyond dreams.

Calling unequivocally for independence, Paine demanded that America decide whether this "king, the greatest enemy this continent hath, or can have, shall tell us '*there shall be no laws but such as I like.*' " The thought of reconciliation was "madness and folly." Britain was beyond forgiveness. " 'Tis time to part." Now.

"Oh ye that love mankind! Ye that dare to oppose not only the tyranny but the tyrant, stand forth! Every spot of the old world is overrun with oppression. Freedom hath been hunted round the globe. Asia and Africa have long expelled her. Europe regards her like a stranger, and England hath given her warning to depart. O! receive the fugitive, and prepare in time an asylum for mankind."

The tone of moral outrage, the resonant anger, and the thrilling call to freedom were irresistible. He forced the colonists to think about and to *rethink* their most closely held beliefs. He forced them to use their common sense.

No one else had ever written about the problem that way. No one wrote like Thomas Paine.

Dickinson, in his farmer letters, could win readers to his viewpoint with a brilliant, marshaled logic and disarming reasonableness. John Adams could hammer a man to his knees with the force of his argument and attack. Jefferson could lead men to the highest realms of morality and human dignity. Sam Adams could raise men's emotions to a killing rage, causing them to reach for their muskets.

But Paine could simplify, explain, then inspire, elevate, exalt, move to noble action. He could unfurl the banner, set the fife and drum sounding, and point to the citadel to conquer.

Sam Adams could make a reader want to kill his neighbor. Paine made his readers want to join with their neighbors in a holy army and go save their country. Paine was thrilling. Irresistible. After Lexington and Concord, he was just what America needed.

"Common Sense" exploded across the colonies. Copies were soon being avidly read in homes and taverns everywhere. Ministers read excerpts from their pulpits. Newspapers cribbed extensive passages from it. Everyone seemed to be able to quote from the pamphlet by heart.

More than anything, "Common Sense" was a great sermon that lifted the hearts of Americans and pointed the only way to a promised land. More than anything before or after, it helped unify the colonists. It "burst from the press," Benjamin Rush wrote, "with an effect which has rarely been produced by types and papers in any age or country." Ben Franklin said its effect was "prodigious."

When he was writing "Common Sense," all through the fall and winter of 1775-1776, Thomas Paine was penniless. Yet he gave the copyright to the Congress, which ultimately earned the royalties on half a million copies.

George Washington, who was struggling against the idea of independence even while commanding the Continental army, was deeply influenced by Paine's writing. "I find 'Common Sense' is working a powerful change in the minds of many men," he wrote. Including his own.

Yet, amid the universal acclamation, and not unexpectedly, there were critics.

John Adams dismissed it angrily as a "poor, ignorant, malicious, short-sighted, crapulous mass." Paine, he said, was "a better hand in pulling down than building." But recognizing its power, he admitted, "I could not have written anything in so manly and striking a style. . . . Thomas Paine's 'Common Sense' is the most brilliant pamphlet written during the American Revolution, and one of the most brilliant pamphlets ever written in the English language."

After "Common Sense," emotionally, intellectually, there was no way back into the British womb. America had been born, and Thomas Paine was its midwife.

A dedicated womanizer, ferocious fighter, and soon to be general, Mad Anthony Wayne was working on his farm in Waynesborough, outside Philadelphia, when he heard the news about Lexington and Concord. He immediately ran for Philadelphia.

It was difficult to say what he was running to—war or women or both. ". . . no prominent American of the times could match [Wayne's] accomplishments in extra-marital affairs. So extensive were his range and conquests that it would be safe to eliminate the qualifying adjective 'prominent' and concede the title.'' Believing that venereal disease was largely confined to the lower classes, Wayne felt he avoided it by never sleeping with any woman who was not his social equal.

The Anglican Reverend William Smith heard the news with profound disquiet. The implications for him as an Anglican minister were enormous. Every Sunday, Anglican ministers were required to mount the pulpit and pray for the well-being of their king and demand from the parishioners total obedience to the Crown. Not to do that was, by law, treason. But in a very few days the most disobedient collection of British subjects in the world would be attending Sunday services in this city hosting radical delegates from all thirteen colonies to the Second Continental Congress. What should he say from the pulpit?

Should he endorse the congress and condemn his king? Or should he pray for his king and condemn the civil disobedience? The noose or the tarpot? His life could be ruined either way he turned. Even more terrifying: Frequently people were whispering a shocking word—Independence. He could no longer be a trimmer, sailing between the two sides in silence.

Reverend Smith was hardly a typical Anglican minister. He knew he had few friends, few people who personally liked him. In fact, the reverend had some serious character flaws. Believed by many to have superb intellectual gifts, he was condemned by others for boundless unministerial greed. At a time when land speculation was almost a communicable disease among Americans, he raised it to an epidemic of self-serving by devoting vast continents of his time to it, pursuing mammon with a passion.

Worse, much worse, he was obnoxiously vain, given to swearing, and when confronted by anyone who disagreed with him, he was openly insulting, sneering, and abusive. In the wake of his dealings with other people, he contemptuously raised up enemies.

Philadelphia's Quakers had never forgotten his attack on their determined pacifism during the French and Indian War, when he excoriated the entire Quaker community as ''a factious cabal, effectively promoting the French interest.'' In the land of the Quakers, he was accusing all Quakers, without exception, of treason.

He had even served time in prison for his wagging tongue until exonerated by a generous court.

In addition, he was a drunk.

And now, just at a time when he needed it, he was discovering that moral courage was not one of his main virtues.

A Scot who had been ordained in England, he had started his ministerial career with great promise. Friends believed him well launched after his thoughts on education—he advocated a combination of practical vocational training and traditional subjects—had captured the attention of practical, traditional Ben Franklin. Franklin brought Smith, even before he had been ordained, to Philadelphia as rector of the Philadelphia Academy and later, in 1755, had him appointed provost of Franklin's new College of Philadelphia, destined to become the University of Pennsylvania.

Politically, Reverend Smith was hardly a blind Loyalist. He had come out strongly against the Stamp Act as "contrary to the . . . inherent rights of Englishmen." Nor was he an extreme radical. In 1774, during the crisis of the Intolerable Acts, he and John Dickinson together reacted to Boston's call for assistance by telling that city, in writing, to pay for the tea. He and Dickinson saw eye to eye on the matter: moderation in all things. State your specific grievances, yes, call for redress, but at the same time express your loyalty. No inflammatory language, no truculent arguments. Be nice.

That was all very fine and, in Philadelphia at least, quite acceptable. Even approved. For nearly a year now, he and six other Anglican ministers of Philadelphia had managed to observe strict public silence over the political controversy that had made the drawing rooms and taverns of the city thunder. But now Reverend Smith saw that the musket shot at Lexington would polarize everything. Very soon he would have to declare what he was—a radical or a Loyalist. He could no longer be the hybrid creature he had been trying to be—a loyal radical or a moderately radical Loyalist. After Lexington, there would be no place to hide.

The Reverend Smith had another problem. His fondest dream might just have gone glimmering. Someone had whispered a warning into John Adams's ear about "the famous Dr. Smith, provost of the Pennsylvania College—a soft, polite, insinuating creature, with ambitions toward an American Episcopate and a pair of lawn sleeves for himself." The Reverend Smith dreamed of being the first Anglican bishop in America. If only . . .

Thus, surrounded by huzzahs of victory, the Reverend Smith hurried through the warm, sun-filled April streets to visit five other Anglican ministers in the city. They were going to draft a shocking letter to London that would make history.

No one in Philadelphia exhibited a stranger reaction to the news from Lexington than Elizabeth Sandwith Drinker. A prominent Quaker, married to the scion of another prominent Quaker family, Henry Drinker, and living in a world of wealth and privilege, Mrs. Drinker, the mother of seven and an inveterate diarist, accepted the role society had thrust on her and, as an obedient housewife and mother, did as was expected of her: She shunned all political controversy, a man's subject. By her own admission she knew little about her husband's business affairs. "I am not," she wrote, "acquainted with the great extent of my husband's great variety of engagements."

Despite the church bells, the marching bands, and the flags, with her entire world about to undergo wrenching changes, Mrs. Drinker's diary for April 24, 1775, is blank. So are the next four days.

On April 29, when more delegates had arrived and there was talk of war everywhere, she noted: "H Walker, E Robinson and M Lever, etc, went away."

And finally, on May 1, with her family of Quakers embroiled in agonizing decisions, the city filling with delegates to the Second Continental Congress, and militia preparing for war by marching up and down in clouds of dust, she wrote: "Bill Bolis, was innoculated by Docr. Redman." Her diary was hermetically sealed from the outside world.

Her views were to change drastically when the radicals exiled her husband and a number of other Quakers to Virginia for their Quaker pacifism. When the British occupied Philadelphia, she was obliged to take a British officer into her home. The officer was named Major Crammond, and he was a late-night carouser who gave noisy parties; commandeered the two front parlors, an upstairs bedroom, and part of her kitchen; and helped himself to her stable of horses. Her Crammond nightmare did not end until the British army left Philadelphia.

Drums, church bells, and marching feet notwithstanding, Sally Logan Fisher, another Quaker and an ardent Loyalist, could not believe that the British army would shoot civilians without cause.

Bitter days were ahead for her family, and she soon discovered personal resources she had never suspected she possessed. Her husband, Thomas Fisher, and a band of other prominent Philadelphia Quakers such as Elizabeth Sandwith Drinker's husband were packed off to exile by patriots. Eight months pregnant with her daughter Hannah when the British arrived, Sally Fisher was almost overwhelmed with the plight of her beloved husband Tommy, and herself and was plunged into depression. The demands of family affairs quickly forced her to deal with the unfamiliar labyrinth of her husband's finances.

She soon had reason to modify her intense enthusiasm for Britain and the British army that occupied Philadelphia under "our beloved General" [Sir William Howe] in 1777. She and all the other American Loyalists in Philadelphia found that they had little more protection from the British army than their Loyalist compatriots in Boston had had. Worse—in Philadelphia, most of the "wanton destruction" and plundering of Loyalists was at the hands not of the revolutionists but of the redcoats themselves. She also discovered that the British officers were not the sterling gentlemen she supposed them to be, and she bitterly condemned their "licentiousness . . . in deluding young Girls."

She was a much different woman in a much different world when her husband returned home in April 1778. She would remain permanently involved in his finances.

The shouting in the street brought home a different truth to the wealthy heiress Elizabeth Graeme Fergusson: Her marriage was about to be destroyed by the Revolution, which would also almost pauperize her. Unlike Elizabeth Drinker and Sally Fisher, Fergusson was a staunch patriot. But her husband, Henry Hugh Fergusson, an immigrant Scot fourteen years her junior and penniless when he married her, was a staunch Loyalist.

The marriage bonds were to audibly snap the moment he returned to Philadelphia from Europe with Lord Howe and the British occupational army. Ignoring her attachment to the patriot cause, the revolutionary committee confiscated her property as part of Henry Fergusson's Loyalist estate. She was to spend many years in the courts before she recovered it.

Elizabeth Fergusson also found that the Revolution was about to bring even greater changes in the role of patriot women. Their participation in political affairs, long forbidden, was now encouraged. They were expected to help enforce the boycott of British goods by not buying them and by pressing other women not to buy them. They were also expected to turn back to their spinning wheels and produce homespun cloth to replace imported British cloth. Spinning became an act of patriotism.

But women went far beyond spinning. They formed their own protest marches and had public confrontations with Loyalist women. Moreover, they crowded onto the premises of any merchant believed to be hoarding. They involved themselves in the war effort in a multitude of ways—they made musket cartridges, sewed uniforms and clothing, baked for the troops, encouraged their men to fight.

Men paid a price for this participation. Women began to state their opinions—even in the columns of newspapers—without apologizing for intruding in the "masculine" world of politics. More than just the Fergusson marriage came to grief over the dinner table when men found their wives differing emphatically, then walking out. Lifelong friendships among women also shattered. Philadelphia women sought to form the first national female organization. Men had to accept these changes. If colonial women failed to support the Revolution—if instead enough of them urged their husbands and sons not to fight—there might not have been much of a Revolution.

At war's end, with women having established their own form of freedom of speech, men could hardly expect the women to return to obedient silence. Yet in this city that was ringing and soon would ring louder during the Second Continental Congress with the words natural rights, the dignity of man, and freedom, none of this applied to women.

No one was more aware of this than Abigail Adams. From her home in Braintree, she would write to her husband in Philadelphia when he was helping to draft the Continental Congress's new law code, urging him to remember to include rights for women. It is significant that she did not ask for the vote. Instead, she wanted the unlimited power of husbands over their wives to be sharply reduced, citing "the power of the vicious and lawless to use us with cruelty and indignity with impunity."

John wrote back: "I cannot but laugh" at "your extraordinary Code of Laws." Then he dodged into dishonest patronizing: "In practice you know We are the subjects. We have only the Name of Masters." In plain English, as he "laughed" he turned her down. There was no recognition of women in the new code.

Thomas Jefferson, who in the Graff house on Market Street, Philadelphia, a year hence would write the Declaration of Independence, which resonates hundreds of years later, failed to see the slavery of his own daughter, newly married, when he prated: ". . . the happiness of your life depends now on the continuing to please a single person . . . to this all other objects must be secondary."

It is interesting to speculate on why women supported the Revolution. Their condition was little changed by it. A woman was owned by her husband. A wife could not sue or be sued; she couldn't write a will, enter a contract, or buy and sell property. Any wages she earned became his property; any property she owned before marriage became his on the wedding day. Usually she did not know the value of the property her husband held, even of the home they lived in, and few knew their husband's income. And she often discovered—with dismay—the extent of her husband's indebtedness only on his deathbed.

Even after the Revolution, the nearest minister could readily silence the complaints of any woman. All he had to do was open the Bible and read aloud God's commandment that Adam should rule over the sinning Eve. Women were inferior to men; women and men had this dinned into their heads from earliest childhood. Predictably, most colonial women had very low self-esteem and low regard for their gender. For most there was no life outside the marriage, outside the home.

Moreau, a French observer of American women in Philadelphia, said: "One can see four hundred young persons, each of whom would certainly be followed on any Paris promenade. These maidens, so charming and adorable at fifteen, unfortunately will be faded at 23, old at 25, decrepit at 40 or 45."

Farmer Josiah Bartlett was winding up his affairs on the family farm in New Hampshire on April 24 so that he could attend the Continental Congress in Philadelphia as a delegate from his colony. His wife, Mary, watched her husband, unaware that her role as submissive wife and housekeeper was about to change permanently. For the first time in her life she was going to step into her absent husband's shoes and take over the operation of the farm.

Writing from Philadelphia during the Continental Congress, Josiah Bartlett sent explicit instructions to Mary on the managing of the affairs of "my farming business." He was to be gone for most of several years, and Mary soon had affairs well in hand and had put so much of herself into the prosperity of the farm that in her letters to Josiah she began, subtly but clearly, to refer to "our farming business."

Susannah Palfrey watched her husband ride off to be paymaster general of the Continental army. Although in his letters to her he warned her against making any unilateral decisions during his absence since he felt she could not "possibly be a Judge" of such matters, he soon trusted her in the most delicate and complex financial affairs, an important lesson for her since he was to die on his way to France in 1780 and she had the whole of his affairs to manage as his widow.

Rebecca Pickering stepped into the business of her husband, Timothy, when he rode off to join the Revolution as part of the quartermaster corps. She quickly revealed a shrewd business sense. Her handling of a complex financial affair that involved the repayment of a debt soon changed Pickering's opinion of her. She became a skillful manager of all his affairs, even supervising the building of their new home while he was away, all the while running a household and caring for their children.

Lucy Flucker Knox refused to be a submissive wife. When she first married Henry Knox, he believed that women were "trifling insignificant animals." So he refused to entrust Lucy with any business matters as he marched off to the war that would cover him with glory as Washington's artillery general. Instead, Knox put his affairs into the hands of his brother Billy. It didn't work.

Lucy must have enjoyed writing to him that the horses Billy had sold for him had fetched only £75, "owing to your not trusting me with the sale of them." Warning him of things to come after the war, she wrote to Knox: "I hope you will not consider yourself as commander-in-chief in your own house—but be convinced . . . that there is such a thing as equal command."

Grace Growden Galloway looked forward to her husband's absence with great relish. Joseph Galloway, still obsessed with his Galloway Plan of Union and avoiding Sam Adams, who was on his way back to Philadelphia, hardly noted her excitement.

As the daughter of Lawrence Growden, justice of the Pennsylvania Supreme Court and possessor of one of the great fortunes of that colony, she had been one of the great marriage prizes of Pennsylvania. She believed that Galloway had married her for her fortune and only that, for afterward he largely ignored her.

Unable to coexist with extremists such as Sam Adams and Patrick Henry, Galloway fled first to British-held New York City, then to England, unaware that he would never see her again. When he ordered Grace to join him and their daughter Betsy in London she refused. Flatly.

She confided in her diary: "Ye Liberty of doing as I please Makes even poverty more agreeable than any time I ever spent since I married." She was even more explicit: "Unkind treatment makes me easey Nay happy not to be with him & if he is safe I want not to be kept so like a slave as he allways Made Me in preventing every wish of my heart."

Grace Galloway had another motive for not leaving Philadelphia. Angered with her husband's mismanagement of some property she had inherited from her father, she wanted to stay and preserve it for Betsy. Her freedom from the husband whose "ill conduct has ruin'd me" became permanent, for Galloway never returned. Their daughter, Betsy, did; she recovered her mother's property from the Supreme Court of Pennsylvania in 1799, more than twenty years later.

In Philadelphia, Mary Norris Dickinson wrote to her husband, John, the Quaker author of "Letters from a Farmer in Pennsylvania." She loved him she said, "with an Affection . . . reaching beyond the present Shadows of a short Existence."

And Mary Logan Fisher wrote to her exiled husband in Virginia: "Oh my beloved, how Ardently, how tenderly how Affectionately, I feel myself thine . . . the impatience I feel to tell thee I am all thy own."

On Fourth and South Streets, the Southwark Theatre, amid all the tumult of the news from Lexington, had no playbill posted, no performance scheduled any time soon.

Even in London, where drama was accepted and encouraged, operating a theater was not an unmixed joy. Troublemakers—who called themselves "Gallery Gods" because they sat in the cheapest, highest seats—were prone to pelting the actors and, by inclusion, the lower audience in most theaters with oranges and pears and garbage.

But theater life in conservative cities such as Philadelphia could be additionally unpleasant. The Quakers, who saw the devil in every line of every play, were constantly badgering acting troupes who toured the city. Often the actors would announce that a performance was a concert or a lecture to misdirect the vigilant Quakers.

The Quakers' antitheater efforts got a great boost when the First Continental Congress, meeting in Philadelphia in the fall of 1774, drafted its Articles of Association which it hoped all thirteen colonies would accept. The articles called for the discouragement of every species of extravagance—including horse racing, gaming, cock-fighting, "shews," and plays.

It must have stuck hard in the craw of David Douglass, manager of an American theater company, to have his acclaimed performances of Shakespeare lumped with cockfights and gambling. Douglass's stature was such that when he was appearing in a theater in Williamsburg, George Washington bought tickets for ten performances and Thomas Jefferson saw one play thirteen times.

Douglass, deciding that the congress's actions were too much, threw in the theatrical towel a few weeks later.

In January 1775, the *Virginia Gazette* published a report from New York City: "The Company of Comedians, with Mr. Douglass, the manager, are preparing to embark for the island of Jamaica, and they will not return to the continent, until its tranquility is restored."

On the day that Israel Bissel galloped into town, Douglass's Southwark Theatre remained firmly locked. And it stayed locked for three more years, until Douglass returned to the city with the British army. General Howe did not share Congress's view on horse racing, gaming, cockfighting, "shews," and plays.

Back in Boston, General Gage, his report to Lord Dartmouth finished, had ordered his messenger, Lieutenant Nunn, to set sail for London on the next available ship. Lieutenant Nunn boarded the commercial brig *Sukey* and sailed with the tide. It was Tuesday, April 25.

That caught Dr. Joseph Warren and the Committee of Safety flat-footed. Working at top speed, the committee would still not have its report ready for some days.

* * *

On April 27, General Gage entered into an agreement with the selectmen of Boston. The general feared that during an attack of the militia upon the city, the civilian residents would join in, confronting him with an inner and an outer enemy. He agreed, therefore, to let those who wished to leave do so and those who wished to enter the city do so—with one stipulation. All firearms must first be stored in Faneuil Hall, properly labeled with the names of the owners. He promised that the weapons would be returned to their owners "at a suitable time." On this day the selectmen received and labeled "1778 fire arms, 634 pistols, 973 bayonets and 38 blunderbusses."

The exodus of people was substantial—thousands, in fact. The ingress—mostly terrified Loyalists—was much less. There were so few Loyalists left. General Gage now feared that with only Loyalists and British military remaining in Boston the rebels would contrive to burn down the town. He later rescinded his order and stopped the general exodus.

On Saturday, April 29, with Captain John Derby on board, the American schooner *Quero* set sail empty, in ballast only, bearing freshly printed copies of the Committee of Safety's report on the battle of Lexington. It was four days after the departure of the brig *Sukey*. The *Quero* hurried after the other ship, hoping to overtake it and make the radicals' version the first report that London would hear. There was nothing else the committee could do at that point but hope for good luck. And that was subject to the notorious vagaries of the sea.

On that same Saturday, at King's Bridge, New York, John Hancock's carriage joined the other Massachusetts delegates. Three miles outside New York City, the entourage was met by a company of colonial grenadiers, a militia company, a long line of chaises and carriages, and thousands of people on foot, clamoring for joy. Hancock said the assemblage raised the greatest cloud of dust he had ever seen.

A mile from the city, a number of men removed Hancock's horses from their traces and planned to pull the carriage themselves into the city. Hancock was delighted, but Samuel Adams objected strongly. He did not want to see fellow human beings reduce themselves to the level of beasts.

New York feted them with abandon. Cheering crowds lined both sides of Broadway, and the New York militia, "well armed and smart in appearance," greeted them.

The delegates, when able to pull free of the adoring crowds, planned to stay in Fraunces Tavern.

There was one mishap in the parade. As the entourage entered the city, John Adams's horse, unaccustomed to the crowd and noise and being driven by an inexperienced young man, took fright and dragged the sulky to flinders. Adams, fully recovered from his brain fever, bought a saddle and was able to complete his journey to Philadelphia on horseback. He was also going to have to replace the sulky, which belonged to his father-in-law.

Within hours of the express rider's arrival in Philadelphia, other express riders were dispatched south and west. One of them headed for Williamsburg, the capital of Virginia, which was in an ominous turmoil of its own.

In Lancaster, west of Philadelphia and the center of the Pennsylvania Dutch country, thoughtful indifference was the typical reaction to news of Lexington. Some of the most fertile farmland in the colonies had drawn Amish, Mennonites, Dunkards, and other refugees from the religious wars of Germany and Switzerland. Frugal, hardworking, and bound to their Bibles and their land, many not speaking English, they felt remote from the issues and reluctant to take sides. After hearing the news, most went silently back to their plows.

Williamsburg

In Alexandria, Virginia, George Washington was happily working on his plantation, Mount Vernon, when the news of Lexington arrived. It was April 26, seven days after the battle. At the time Washington had as a houseguest the quixotic, fascinating Charles Lee, a retired British military officer.

Washington quickly arranged his trip to Philadelphia and the Second Continental Congress. As an indication of how he was thinking, he carried with him his striking uniform of the Virginia militia. In reaction to recent events, Washington wanted his uniform to visibly state his position to the delegates: He was ready to fight. Some in Philadelphia would interpret it as an application for the job of commander in chief.

To Charles Lee, bored in his retirement, the news from Massachusetts suggested a new career opportunity.

Although Washington and Lee went by different routes, both were in Philadelphia when the second session began—seeking the same military assignment.

Charles Lee presented a strange mixture of slovenliness and brilliance. Rising to second in command of the Continental army, he was to end a brilliant career by destroying it, then die in deep financial straits.

Born in Chester, England, Lee was less than a month older than Washington. A scholar and a soldier, he received an excellent classical education in England and Switzerland when, in 1748, he followed his father into the army. Lee was fluent in Latin, Greek, Italian, and French, and wherever he went—as a traveler or a soldier—he carried his library—which included his three favorite authors: Shakespeare, Plutarch, and Jean Jacques Rousseau.

A brilliant satirical writer and a gifted, sharp-tongued speaker, he was plagued with an uncontrollable temper and a foul mouth. A collection of contradictions, he wore expensive custom-made uniforms, which in time became food stained and completely rumpled after nights of sleeping in them.

Lee had been nicknamed "Naso" behind his back because of his large nose, which projected from a bony, unattractive face. He was five feet nine—tall in colonial days—and slender, with noticeably small hands and feet. With age, he would suffer inside a gouty, arthritic body. He was exceptionally bright, passionate in his opinions, sharp-tongued, and, during his mood swings, witty in conversation. Outside the military, he proved to be a brilliant propagandist and writer and a charismatic speaker.

His adoring pack of dogs went with him everywhere: on the road, into battle, and even into his bed. He said he preferred them to humans.

He was not the kind of man women were attracted to, and despite rumors that he had married the daughter of a Seneca chief, he was destined to die alone, a bachelor. Yet observers said he "was a genius at getting on with the Indians; they called him Boiling Water because he was never still."

Fearless in quarrels, fearless in battle, and always a dangerous adversary, he was considered by many who witnessed his wild tirades to be mentally deranged. Others found him so obnoxious and self-serving that they simply avoided him. In arguments, unfortunately for his opponents, he proved to be too often correct. He was "an odd genius."

Beginning his military service in Ireland, he later, at age twenty, served as a lieutenant under General Braddock and was present at his death during the ambush and defeat at Fort Pitt in the French and Indian War—along with General Gage and George Washington. In 1758, Lee was wounded leading a futile attempt to storm Ticonderoga, which cost him two fingers. While in America, apparently, Lee conceived a romantic notion of the American as a new breed of man who could develop unconventional military tactics to create an ideal utopia, free of European influence.

He won some glory in Portugal in 1762 under General John Burgoyne when his cavalry routed a Spanish camp, and he was given the rank of colonel by the Portuguese army.

After a grand tour of Europe, Lee returned home. Believing that King George III had broken a promise of military promotion, he allied himself with the Whig Party and became friends with two firebrands—Isaac Barré and John Wilkes—and their associates and was soon using his acid pen and powerful public speaking voice to attack the Tory Party. Rumor had it he was Junius, author of the notorious London letters to the king.

The whole world seemed to know about Charles Lee and about the pack of hound dogs that accompanied him everywhere. Above all men, it was said, he could whip an undisciplined army into shape.

Due to his speeches in defense of the colonies, he became much in demand in America. On October 8, 1773, still on half pay as a lieutenant colonel in the British service, Lee sailed to New York and walked into the welcoming arms of all the major radicals in the colonies—including Samuel and John Adams, Richard Henry Lee, Robert Morris, and Charles Thomson.

Lee purchased an estate in Virginia and promptly busied himself with the rebels' cause. He was now forty-two years old and in his prime. He would never leave America again.

Addressing himself to one of the radicals' greatest fears, Lee attracted widespread attention when, as the leading military expert in America, he assured the colonists they could stand up to the British army and win. When he publicly called for American independence—he was one of the first to do so—he was saluted as the Palladium of American Liberty.

Traveling to the Second Congress in Philadelphia, he expected his stature to gain him command of the revolutionary army.

At issue here was more than just the conflict of two casual friends becoming military rivals: Lee and Washington represented two exactly opposite military philosophies.

Lee believed that in any future war the American soldier was too freewheeling and intelligent to be forced into the traditional European army format but should rather fight in an American style. Today, Lee's tactical concepts would be called hit-and-run guerrilla warfare: harassing the enemy, disrupting his supply lines, inflicting casualties until they became unacceptable, never fighting in the open, never being maneuvered into major engagements. His views had the ear of a number of American leaders.

General Heath's ring of fire during the Lexington campaign was exactly what Lee had in mind.

Washington rejected Lee's idea of partisan warfare. He felt that for America to win respect in European capitals, a trained European-style standing American army would have to convincingly defeat the British on the battlefield in traditional warfare.

Congress would have to decide which form of fighting would be used.

When the Second Congress considered Lee's application to be commanding general, both his English background and his imperious personality held many congressional delegates back, but in the end, mainly at Washington's urging, the congress granted him a major general's commission, third in command after George Washington and Artemus Ward. In time, he would become Washington's second in command. But he was never first.

The congress, in addition, compensated him for the lifetime half pay he would lose when he resigned from the British army—as well as for the expected loss of his English estates, which the king would surely confiscate—by granting him 30,000 Spanish dollars to purchase his Virginia estate and settle his debts.

For the time being, the congress had accepted Washington's view of battle. Lee would accompany Washington to Boston and the battlefield where their two opposing military philosophies would be tested.

In the end, Lee would prove to be as much thorn as rose on Washington's staff. Beside Washington's upright, immaculate behavior, Lee's conduct was often disreputable and shabby. While he publicly scorned General Howe because he "shut his eyes, fought his battles, drank his bottle and had his little Whore," Lee's conduct was little better.

When he spent the night at Washington's headquarters in Valley Forge, Washington gave up the best bedroom at the back of the house to him, moving himself and Martha to the front bedroom.

Colonel Elias Boudinot of New Jersey described the scene the next morning: ". . . he lay very late and breakfast was detained for him. When he came out, he looked dirty, as if he had been in the street all night. Soon after I discovered that he had brought a miserably dirty hussy with him from Philadelphia (a British sergeant's wife) and had actually taken her into his room by a back door, and she had slept with him that night."

Within a few weeks, at Monmouth, he would be suspected of treason.

On April 27, the news of the fighting in Massachusetts reached Annapolis, Maryland, and the ears of one of the most pugnacious Anglican ministers in all the colonies. The Reverend Jonathan Boucher of Queen Anne's Anglican parish was famous for his habit of giving Sunday sermons with two loaded pistols on the cushion beside his Bible.

An uncompromising Loyalist, Boucher did not hesitate to upbraid, excoriate, vilify, and condemn his anti-Loyalist flock for their mental sins and sick opinions.

His enemies knew he was not a man to be baited casually, for he was known far and wide for having beaten senseless the village blacksmith. The people in Annapolis realized that his standoff with the local radicals could not last much longer, and they watched and waited for the final showdown.

From near poverty—his father was a penniless English schoolteacher who also operated an alehouse—the thirty-seven-year-old Boucher, educated by his father, came to America to become a tutor of the children of a Virginia planter. With additional education, he was ordained an Anglican minister. In time, he came to own a large plantation with a number of slaves as well as a boarding school for thirty or forty boys, who included George Washington's stepson, Jackie Custis.

The reverend was also very active in colonial politics. A member of the Maryland Assembly, he wrote or edited most of the bills presented, most of the speeches and addresses of the governor, as well as the major papers of the governor's council. He was dead set against any jumped-up, unlettered folk who wanted to help run things.

An active polemicist in the colonial newspapers, he was also a patron of the theater, wrote songs and epigrams, and somehow in the midst of all this managed to find time also to serve as minister in parishes in both Virginia and Maryland. But more than anything, he battled every Presbyterian and Baptist minister within reach, endlessly extolling the great virtues of the Church of England. He was right. They were wrong. And if they offered any physicial objection while he was preaching they would be dead wrong.

Without hesitation, even in this center of slavery, from his pulpit he condemned slavery and even the treatment of freed slaves. "An African slave," he preached, "even when made free and supposing him to be possessed of talents and of virtue, can never in these colonies be quite on terms of equality with a free white man."

He did not release any of his own slaves—Virginia law forbade that, although had he done so it would have been an act of sincerity that would have lent still greater weight to his words. Instead, in the eyes of the planters, he did something far worse. He taught his slaves to read—a piece of madness worse than teaching women to read.

Starting with several of his own slaves, Boucher taught them their letters, then with their assistance conducted Sunday afternoon instruction classes in religion to large groups of black slaves, sometimes as many as 1,000. He baptized hundreds and then stunned the whole colony by bringing thirteen of them into his parish services in St. Mary's Church, in Caroline County, Virginia.

And in the midst of passionate, violent anti-Catholicism, Boucher defended the Catholics' right to freedom of conscience. The planters were scandalized.

It was just a matter of time. . . .

When Boucher heard the news of Lexington and Concord, he was struck with great dismay. His adopted country was lurching toward war. Crossing the Potomac River heading south, he encountered midstream his friend George Washington, who was traveling the opposite way. Boucher warned Washington that his trip to the Second Continental Congress would lead to civil war and a campaign for independence.

Washington was greatly shocked. "If ever you hear of my joining in any such measures, you have leave to set me down for everything wicked," he said.

The issue came to a head a few months later, in July, when Boucher was confronted in his church by the Annapolis radicals. It was the Day of Fasting and Prayer, ordained by the Second Continental Congress. Reverend Boucher, sermon in hand, a brace of pistols in his belt, opened the door of his church and found it stuffed with men—200 of them—along with their leader, one Osborne Sprigg, sitting in the pews waiting for him.

Acting on the specific request of Governor Eden of Maryland, an avowed Tory, the Reverend Boucher had gone to the church prepared to give a sermon against the Continental Congress to these dangerous men—who were determined not to hear it. Sprigg had given unequivocal orders to twenty men: The moment Reverend Boucher set foot on the steps of his pulpit, he was to be shot dead.

With pistol in hand, the Reverend Boucher began to ascend the steps to his pulpit. The church erupted into a scuffle. Boucher seized Sprigg by the collar, placed his cocked pistol against Sprigg's temple, and said, "Sprigg, if any violence is offered me, I will instantly blow your brains out."

Then, still leading Sprigg by the collar, the reverend exited his church, let go of Sprigg at last, mounted his horse, and rode away. The theme Boucher had chosen for that special day's sermon against the Continental Congress? Peacefulness.

Still beyond intimidation, Boucher returned the following Sunday, outshouted the Whigs, faced down their weapons, and gave his sermon. Haranguing them into silence, he roundly damned their participation in what he deemed an immoral, totally unjustified rebellion.

But by January 1776, some five months later, he had lost the argument. Cast out of his pulpit and school, fearing for the safety of his family, he was back in England, unrepentant, convinced to the day he died that the Whigs were wrong. "In America," he said with seething scorn, "literally and truly, all power flows from the people."

On April 27 in the tidewater section of Virginia, on the road to Philadelphia, Peyton Randolph, the fifty-four-year-old speaker of the House of Burgesses, Virginia's legislative body, was traveling from his home in Williamsburg by coach when he learned of Lexington from an express rider. Randolph was a planter, major slave owner, and lawyer; a cousin to Thomas Jefferson, and one of the most powerful politicians in his colony.

Wealthy, intelligent, and superbly well educated, Randolph was hardly a typical revolutionary. Born to one of the oldest aristocratic families in Virginia, he was a graduate of William and Mary College and had studied law in the Inner Temple in London. He was also president of both the First and the Second Continental Congresses.

He had led the Virginia protest of the Stamp Act—a litmus test for all patriots—and was now chairman of the Virginia Committee of Correspondence, one of an intercolonial network of radical committees created to exchange information, news, and techniques for opposing the Crown—all of which put him on the British cabinet's enemies list.

Peyton Randolph was also the leader of tidewater planters who were struggling to keep control of the House of Burgesses in the face of growing opposition from Patrick Henry and small farmers from the western counties of Virginia. This was a simple struggle for power—old money versus broken fingernails, large plantations versus small holdings, Anglicans versus Presbyterians and other dissenter groups. One of the continuing issues was the planters' increasing use of slaves. Under Patrick Henry's leadership, the new and the old money had begun to share power. Temporarily, the two opponents had tabled their differences while they coped with a common antagonist, Royal Governor Dunmore and the Crown he represented.

Randolph presented himself as a moderate, was actually a conservative, and was slowly becoming a radical. Left to ponder the significance of Lexington as he rode north, he would find himself becoming even more radicalized in Philadelphia.

With Lexington as a goad, he picked up the pace a bit.

Late at night, under the wide and starry sky of April, the sounds of pounding hoofbeats and a shouting voice woke many residents in Williamsburg who had long ago gone to bed. The express rider seemingly burst into the darkened city, calling out the news of fighting at Lexington. It was Friday, April 28. In the handwritten copy of the original handwritten note from Watertown, Israel Bissel's name had become Trial Brisset. Soon everyone was up and talking in knots. Shortly other riders would carry the news down rural roads to plantation houses and villages.

As the capital of Virginia, Williamsburg was the social and cultural center of the colony. With no industry, it had a year-round population of barely 1,600, half of which were slaves. Normally it was little more than a village—dull, sound asleep. But twice a year, in spring and autumn during Public Times, the sitting of the House of Burgesses, the city leaped into life.

Then the population would soar to 6,000. People would flock to see the legislative confrontations between governor and burgesses, to do business with government, or to meet and make business deals with each other and attend tobacco auctions and slave auctions.

Planters and their entire families came to socialize with other plantation families. Arrange marriages. Gossip. Play cards. See—and be seen in—the latest London finery. Attend balls. Dance until dawn. Participate in the academic life of William and Mary. Attend the British repertory theater, featuring everything from farces to Shakespeare. Williamsburg's many taverns would be choked with loud laughter, drink, tobacco smoke, and talk. Pigs would be roasted on spits in kitchen fireplaces. There would not be a spare bed in the whole town. When the House of Burgesses was in session, provincial Williamsburg dropped its polite manners and became rude, crude, dusty, noisy, drunken, even unpleasant. But for the sociable, Public Times were not to be missed.

When the express rider cantered into Williamsburg that night, the town should have been wide awake and roaring with Public Times. But not this spring. The town remained a somnolent village. Royal Governor Dunmore had angrily closed the House of Burgesses nearly a year before when the burgesses had declared a Day of Fasting and Prayer in protest of the closing of the port of Boston. He was still refusing to open it.

Inside Williamsburg's Royal Palace, the governor himself had been hastily packing, preparing to flee. When servants came running up the stairs with the news from Lexington, flight became even more urgent. Early the next morning, before most of the town was stirring, he had his wife and children trotted off in the royal carriage to the British warship HMS *Fowey*.

The governor himself would not be far behind. Dunmore's nemesis, Patrick Henry, was on the road, marching a troop of armed men on Williamsburg, on *him,* and Dunmore had no intention of being there when Henry arrived.

The governor felt he had no choice but to leave. Casting about for weapons, any weapons, the biggest defense he could muster against Patrick Henry would have been about forty British seamen. Had he a brigade of regulars, Dunmore would have attacked with fury, and the American Revolution might have started in Williamsburg—Dunmore versus Patrick Henry—even before Lexington.

Henry had raised an army against the governor because Lord Dunmore, copying General Gage, had snatched from Williamsburg's town arsenal its entire supply of gunpowder and locked it away on board the HMS *Fowey*. Unfortunately, the governor had not prepared

Patrick Henry

himself for the colonists' reaction. To many Virginians, confiscating their gunpowder was in itself an incitement to war.

That same night, John Pinkney, the editor of *Pinkney's Williamsburg Gazette,* was living through the most turbulent night in the history of his newspaper.

Trying to put his journal to bed, he found himself awash with news—more than he had space for. For his normal edition of Saturday, April 29, he had started by furiously setting type on the tumultuous Dunmore gunpowder affair, using lots of exclamation points.

While he was in the midst of that he received equally disturbing news from London. Lord North was seeking authority from Parliament to extend the New England Restraining Act to all the other colonies. If passed, it meant martial law in Virginia.

In itself this was shocking news, but what made it even more so was that the colony had been led to expect something much more appeasing from Lord North—a "grand conciliating plan," not a mailed fist. (News that this alarming bill had been defeated by both Houses of Parliament did not come until later.)

The frantic Pinkney started to reformat his paper. Then he received yet another shocking report—this one from New York. General Gage had received authority to capture, try and execute Sam Adams, John Adams, John Hancock, and—unexpectedly—Peyton Randolph.

Now overwhelmed, Pinkney decided he had to publish a special supplement. But even the added pages barely accommodated all the typeset. Then came the hoofbeats that echoed those in Lexington. Americans and British soldiers had battled each other. That could mean war. But Pinkney simply had no room for any more news—good or bad—and no time to redo the whole edition.

So, to make his printing deadline, he hastily hand-pegged type on the story from Lexington and stuffed it—a very short item—on the back page, right next to a public notice from George Washington. Washington was offering a reward for the capture of "two runaway servants, a joiner and a brickmaker."

And that was the best Pinkney could do. He began the slow process of hand printing his newspaper, determined to meet his deadline. Things were getting quite out of hand.

Meanwhile, Patrick Henry was marching ever closer—reaching with both hands for Dunmore and the gunpowder.

For over a year, Governor Dunmore and Patrick Henry had been circling each other, making threatening gestures. Possibly, without Henry, the issues could have been resolved at least temporarily. But Henry was in no mood to temporize, and he kept the colonists in a constant state of emotional turmoil. As for Governor Dunmore—he was the last man the king should have sent to Virginia. The aggressive Scotsman was not given to ruling passively. This made a clash between the two inevitable.

Dunmore's situation was not unique. The governors of all the colonies were living lives of constant harassment. The job was unpleasant at best, and lately so unpleasant that there was an unprecedented turnover of governorships. England was now home to three ex-governors of Massachusetts, three of New York, four of New Jersey, four of South Carolina, and three of Virginia.

Dunmore's unimpressive set of credentials had made him unwelcome in Virginia even before he arrived. He had been governor of

New York, where he "had found himself some congenial friends (too congenial, others thought), and he set about quickly acquiring a considerable estate."

James Rivington's *New York Gazeteer,* upon reporting that Governor Dunmore was being transferred from New York to Virginia, said, "He is a Chearfull free liver & an Anguish Climate will ill suit his convivial Disposition." Another source accused him of being a "drunken rake" —one who was reluctant to leave New York and reluctant to serve in Virginia.

Worse, he was accused of neglecting his wife. One commentator wrote in his diary, "his Lordship is I believe a Consummate Rake & does not pay that attention to his Lady that she seems to deserve." "His wife proved to be so popular," another writer noted, "she deserved a better husband."

Worse than that, Lord Dunmore was a Scot.

Worst of all, a Presbyterian.

And in addition to everything else, Lord Dunmore was accused of being land hungry. That wasn't unusual. In Virginia it might even have been regarded as a virtue. Almost every man in the colony, including the highly respected Colonel George Washington himself, was avid in the acquisition of land. But Dunmore was an outsider.

When he arrived in Virginia, Earl Dunmore promptly lived up to his billing. Soon after taking up his duties as governor, the earl bought a residence called Porto Bello sitting on some 579 acres of land about six miles from Williamsburg. He owned another 2,600 acres in Berkeley County, plus another 3,465 in Hampshire County. His net worth also included fifty-six slaves worth about £6,000 and a dozen indentured servants. He also valued his furnishings and other property inside the governor's palace in Williamsburg at £3,200.

Critics accused him of starting "Lord Dunmore's War" in order to acquire still more land. They said that this armed foray into the wilderness beyond the Virginia mountains, ostensibly to survey the terrain and clear up a conflict over borders with Pennsylvania, was nothing more than a thinly disguised attempt to snatch up a tract of heavily timbered land from the Indians.

It wouldn't be long, however, before Dunmore's troubles with Patrick Henry distracted him from his land quests. The power struggle that ensued was between two extraordinarily stubborn men, neither of whom would go away or back down.

Their most recent confrontation had started a few weeks before when Patrick Henry had thrown a thunderbolt directly at Dunmore's head from the extralegal Second Virginia Convention.

In "one of the greatest speeches in the annals of American oratory" Henry stunned Virginia, shocked the other colonies, and terrified Dunmore.

In attendance was every one in Virginia who mattered—Peyton Randolph (who was chosen president), Jefferson, Washington, Edmund Pendleton, Benjamin Harrison, Francis Lightfoot Lee, Richard Henry Lee, and a host of others, the leading landed gentry and the real power of Virginia, meeting in Saint John's Church in Richmond. The purpose of the gathering was to ratify the equally illegal acts passed by the First Continental Congress and to elect delegates to the Second Continental Congress in May. To the attendees, it seemed at the time to be a simple event—three days in which to approve, with some gentlemanly debate, the acts, pick the delegates, talk a little treason, and then return home. But that was before Henry rose to speak.

Most of the men were comfortable planters who were uncomfortable defying their king. All they really wanted was to have the king put things back together again the way they had been, get Parliament off their backs, make some changes in mercantilist policies, and let them get on with their planting. And they were hopeful that the Crown would soon do just that. But that's not what Patrick Henry wanted—or expected. Up on his feet, where he was always unpredictable and dangerous, in short order he made their planterly mouths fall open. He demanded that Virginia raise a militia. It was all Randolph could do to keep order as others clamored to speak against Henry.

For these were sensitive times: Virginia had friends in London who were pressuring the king to redress grievances. Raising a militia would make those friends back off, make the king even more intransigent. British merchants were urging Parliament not to cause another American boycott because the first one had hurt them severely. People in power in London were listening. Ideas were being exchanged. Reconciliation was expected to arrive on the next ship from England, or perhaps the ship after that—soon, anyway. Lord North was said to be preparing a plan to smooth things over. Besides, the idea of Virginia's raising a militia to fight the Crown—the mightiest power on earth—was unthinkable to these men.

But Henry couldn't be shamed or silenced. Much to the annoyance of many there that day, Henry took the podium once more:

> Reconciliation? What reconciliation? We are infested with armies and fleets sent from England. A military presence with only one purpose—to force our submission.
>
> And what is our reply? More argument and petitions? Sir, we have been trying that for the last ten years. We have nothing but the same old arguments to present. We have already done enough petitions, remonstrations, protestations.
>
> There is no longer any room for hope. If we wish to be free—if we mean to preserve inviolate those inestimable privileges for which we have been so long contending . . . we must fight! I repeat it, sir, we must fight! An appeal to arms and to the God of Hosts, is all that is left us!
>
> They say we are weak. But when shall we be stronger? Next week, or next year? Shall we wait until there's an army stationed here, a guard at every door? No.
>
> We are not weak. We are invincible in the holy cause of liberty. Three million people invincible by any force which our enemy can send against us.
>
> There is no retreat, but in submission and slavery! Our chains are forged. Their clanking may be heard on the plains of Boston! The war is inevitable—and let it come! I repeat, sir, let it come!
>
> Gentlemen may cry, peace, peace—but there is no peace. The war is actually begun! The next gale that sweeps from the north, will bring to our ears the clash of resounding arms! Our brethren are already in the field! Why stand we here idle? What is it that gentlemen wish? What would they have? Is life so dear, or peace so sweet, as to be purchased at the price of chains, and slavery? Forbid it, Almighty God! I know not what course others may take; but as for me, give me liberty, or give me death!

The shocked silence, then the standing ovation, signaled that Patrick Henry had moved many.

The Second Virginia Convention went on to approve the acts of the Philadelphia Congress, including the embargo, to be effective September 1. And it chose seven delegates to the Second Continental Congress, including Randolph, Washington, and Henry. Jefferson was chosen to stand in for Randolph should the latter not be able to attend.

Yet, even with Henry's words of war pealing in their ears, the majority of delegates still hoped for reconciliation with the Crown.

At adjournment, Patrick Henry had to feel victorious. He could claim that his speech "had whipped up a gathering of quietly happy Englishmen into a band of militant Americans." Maybe not. But most important, Henry got his militia. Before adjourning, the convention called upon each county in the colony to raise at least one company of infantry and one troop of cavalry.

After these proceedings, Governor Dunmore was more isolated in his palace than ever, powerless and under siege. He needed to find some weapons to fight with. He turned his eyes to the powder stored in the city's magazine.

The previous March 28, less than a month before Lexington, and three days after the Second Virginia Convention had adjourned, Lord Dunmore issued his official reaction. On orders from the secretary of state in London, he said, he forbade such gatherings as the Virginia Convention or the Philadelphia Congress. Specifically, he wanted to stop the sending of delegates to Philadelphia. The delegates, outraged by this intrusion of British authority into colonial affairs, ignored the Crown's unenforceable order.

Meanwhile, in a much more provocative move, each Virginia county was raising a militia unit. As in Massachusetts, men were marching and training with arms throughout the colony. Virginia seemed to be on a war footing, especially to the governor. To combat force with force, Lord Dunmore would have needed a brigade of redcoats. He had none. But to avoid losing complete control of the situation, he decided to take some kind of action.

In a sudden unforeseen strategem, he sent a band of seamen to the public magazine in Williamsburg to seize the colony's gunpowder—some twenty-one and a half barrels—which were placed on board the *Fowey*. Virginia had not yet learned of Lexington.

The governor justified his move by saying he expected an uprising of slaves who would seek to arm themselves with the powder and muskets. The town demanded the gunpowder back on the same grounds—fear of a slave uprising, against which the planters needed the powder to protect themselves and their families.

For slave owners, an uprising was always a terrifying prospect. The slave population in Virginia stood at 140,000—twice the white population. In fact, Virginia had the largest slave population of any of the colonies, 100,000 more than Pennsylvania or Massachusetts.

And through births and additional slave imports, the slave population was increasing.

Dunmore assured the colonists that in the event of a slave uprising he could have the powder back in the magazine in half an hour, but no one believed him. They insisted on having the powder back. Immediately. He refused.

Radicals in Fredericksburg gathered an army. They prepared to march on Williamsburg and the governor's palace. Dunmore, in a rage compounded by fear, publicly threatened that if troops approached he would emancipate the slaves, arm them, and burn Williamsburg to the ground. There was no more incendiary act a governor in Virginia could have threatened. The whole countryside was aghast.

The next day, the troops at Fredericksburg held an urgent meeting. After much discussion, the troops, fearing that Dunmore would make good on his threat, dispersed.

Now that he had laid his hand on this potent but exceedingly dangerous weapon, Dunmore was ready to brandish it anytime he was pressed. He felt it had finally given him the upper hand that could wring still more concessions from the colonists. He hadn't counted on his nemesis, though, who was supposed to have already left for Philadelphia and the Second Congress.

But Henry was no cautious planter. Completely unimpressed by Dunmore's wild threats and eager for a fight, he put off his departure and, instead, gathered a party of frontiersmen from his native Hanover County to march on the capital.

Although he often liked to pretend to be an unlettered country rustic, Henry was hardly a frontiersman himself. His father was a judge of the Hanover County court, as well as a county surveyor and a colonel in the Virginia militia. And Henry himself had an excellent education—from his father, at home, and also from his uncle Patrick, who taught him Latin, Greek, and the Bible.

In later years, Henry was to be widely known for rarely reading anything, but in youth he consumed books by the dozen, especially history, from which he could snatch apt examples and quotations. People soon discovered that this adolescent with his red hair that early subsided into a fringe and his excitable blue eyes was also an eager talker and a debater. From earliest youth, Patrick Henry liked a good fight.

At eighteen, dirt poor, he married Sarah Shelton, who was also dirt poor. He attempted to make a go of a small farm with six or seven slaves, then watched powerlessly as the farm failed and his farmhouse burned down. He set up a store, failed at that, started a second, and failed again. Now with children, he was reduced to working in his father-in-law's tavern, living on the floor above. With time on his hands, he turned to the study of law. It was a most improbable legal scholarship.

The sons of the wealthy in Virginia studied law in the Inns of Court, in London. Others would apprentice themselves to a lawyer in Virginia for months, even years. Thomas Jefferson had prepared for the law by attending William and Mary College in Williamsburg then serving a five-year apprenticeship in the law office of the brilliant George Wythe, professor of law at the college and one of the state's leading law scholars.

Henry had no patience with that. He acquired a copy of a well-known text—*Coke on Littleton*—and invested six weeks of his time reading it. On the strength of reading that one text, he then journeyed to Williamsburg and asked the legal examiners to certify him as a lawyer.

He couldn't have picked a more august body of four examiners. Peyton Randolph had studied in the Inner Temple in London, his brother John had studied law at Grey's Inn there. Robert Carter Nicholas, who came from great wealth, was recognized as the head of the Virginia bar. The fourth was the formidable George Wythe. Henry needed two signatures.

Wythe dismissed him after a few questions.

Robert Carter Nicholas, noted for his kindness, hesitated. Henry pressed. Nicholas hesitated. Henry knew so little law. Henry pressed some more. Nicholas extracted a promise that Henry would study more law. Then he signed.

John Randolph, during his interview, thought that Henry had already gotten the two required signatures. He enjoyed—even admired—Henry's debating skills, was impressed with his general education, and scored him zero on the law. He also thought that Henry was a youth of genius. He signed.

Peyton Randolph was equally impressed—with the man but not his knowledge of the law. He, too, used the word *genius.* He, too, signed.

Patrick Henry, the most ill-prepared lawyer in Virginia, had his license. He was twenty-three.

George Wythe

He quickly built a practice and, representing the poor and the unsophisticated, learned the law the hard way—in the courts fighting other more knowledgeable lawyers. Finally, in the case of the Parson's Cause he was given the opportunity to gain the fame he yearned for.

In a cash-poor society, Virginia planters paid their parsons in tobacco, 16,000 pounds a year. That amount was set by royal law in London. But when the value of 16,000 pounds of tobacco reached 400 pounds sterling—which it finally did in a rising market—the planters began paying in paper currency—at the rate of two pence a pound. The parsons were shortchanged by two thirds every year.

With fourteen mouths to feed, the Reverend James Maury was threadbare and hungry. He needed full payment desperately. He sued. The court sympathized. It told the vestry of Maury's church that the law was the law, that it was illegally holding two thirds of Maury's income, and to pay up.

All that remained was a hearing to set damages. Reverend Maury's lawyer, the ponderously heavy and widely respected Peter Lyons—with 300 pounds of charm, loquaciousness, and skill, walked into court sure he had an easy case. He was suing not the church but Deputy Sheriff Thomas Johnson, who was looking at a staggering fine for not collecting the full value of the tobacco owed to Maury. Opposing him, at a fee of fifteen shillings, was Patrick Henry, with the ink on his law certificate still wet. He got the case because no one else wanted to touch it. And because Johnson had no money.

Instinctively, Patrick Henry did what Sam Adams would have done. He ignored the law, the legal precedents, the cut-and-dried court case that said Johnson was going to pay. Instead, Henry looked for and found the emotional issue. He astonished the court and the jury by attacking the king of England.

This whole case, he said, was the fault of the king for not approving the Two Penny Act, passed by the Virginia House of Burgesses, which was proper legislation, properly enacted, and which would have properly regulated this whole matter. The king, said Henry, had thereby forfeited his claims to obedience. Shouts of treason filled the courtroom. If Henry's argument was short on logic, it was loaded with emotion—and the jury was listening to every hypnotic word. He was saying the law was not valid; therefore, king or no king, the law need not be obeyed.

Henry then went on to attack the Anglican clergy. Among the jury were dissenters forced to pay support to the Anglican religion they hated. Young Henry's rhetoric soon had the courtroom verging on applause with his every statement.

In the end, the poor parson was awarded his damages—one penny.

Henry afterward apologized to the parson for attacking him, but the name Patrick Henry became famous throughout the colony. The poor had found a new, powerful voice. His law practice grew accordingly. To the planters, his reputation as a dangerous man also grew.

At the age of twenty-nine, Henry got himself elected to the House of Burgesses, just as the Stamp Act was enacted into law. After listening to the planters exclaim against the act in their usual mild-mannered way, he promptly tore it to shreds, found moral outrage in every word, saw the deepest and darkest Parliamentary conspiracy in every paragraph. Henry was not the type of Virginian who would rant and rave, then sit down. He wanted action. To correct matters, he introduced his defiant seven resolves. To the shocked burgesses, Henry's demands sounded like a call to war.

His timing was perfect. The spring session was at an end, and most of the older legislators, thinking major business was finished, had left for their plantations. Most of those remaining were young hawks.

The most important of Henry's resolves was the fifth, which read: "Resolved, That the General Assembly of this colony have the only and sole exclusive right and power to lay Taxes and Impositions upon the Inhabitants of this Colony . . ." There it was in writing: Parliament had no authority to tax Virginia.

And even more incendiary were the sixth and seventh resolves. Clearly seditious, some said. The resolves emphatically asserted that Virginians were "not bound to yield obedience to any law or ordinance whatever . . . other than laws or ordinances of the General Assembly." That was a flat-out declaration of independence from Parliament, from the Crown, from England.

The older and wiser heads—the two Randolphs, Peyton and John, and George Wythe—attempted to delay debate on the Henry resolutions until they heard from England on earlier ones.

They were overridden. To contain this wild man from the hills, they removed the sixth and seventh resolves and rephrased the others. In the end, despite the best efforts of the conservatives, the modified but still defiant Henry Bill was passed. It was a stunning piece of legislation. With these five Henry resolves, Virginia said that the Stamp Act was illegal, not to be obeyed. Implicit was the Virginia offer to settle the matter with London on the battlefield.

But something of equal significance had happened. Red-clay Virginians Patrick Henry and Richard Henry Lee had also confronted the tidewater planters who controlled office and largess in the House of Burgesses and bearded them in their seat of authority. This was an internal power struggle, and the balance had just swung to west Virginian counties. With the passage of Henry's resolves, the plebian radicals would lead Virginia in the fight against England. Historians mark the Henry resolves as the beginning of the revolutionary movement in the colonies.

When Henry's Virginia Resolves were published and circulated among the other colonies, eight of them, believing the House of Burgesses had adopted all seven, also adopted all seven. Connecticut accepted the Virginia version with barely a word changed.

Henry had beaten Sam Adams to the punch—he was the first to oppose the Stamp Act and soon enjoyed an equal amount of fame. He became a ranking member of the Virginia revolutionary Committee of Correspondence. Attacking king and Parliament and royal governor at every turn, he was Virginia's most pugnacious radical. His brilliant eloquence would draw a crowd any time he rose to speak. And he spoke often. Thus, the rivalry between the two men to lead America to independence began.

The night the express rider shouted the news of Lexington along the streets of Williamsburg, Patrick Henry's army camped by Doncastle's Ordinary, some sixteen miles from Williamsburg. Henry, showing not even a hint of backing down, sent his demands to Dunmore: Return the gunpowder or pay for it. He had called Dunmore's bluff. For, despite his threats, Dunmore had hesitated to issue a proclamation freeing the slaves. Even he could see the unpredictable volatility of that act.

After complex negotiations, Henry was given a draft from the colony's treasurer to cover the cost of the gunpowder.

Dunmore had backed down, if only for the moment. Henry's irregular army disbanded, and Henry himself prepared to travel to the Continental Congress in Philadelphia. Dunmore returned to his palace. There he issued a proclamation declaring Patrick Henry an outlaw for disturbing the peace.

The colony returned to its vitally important springtime business of planting corn, cotton, and tobacco. People seemed to have lost interest in Dunmore's posturing. Even the justice system returned to its mundane assignments. Four prisoners—"Messrs. Pitman, Gray, Wood, and Watkins"—were brought up on charges: the murder of a black boy, two robberies, and a rape. The four were all found guilty, and all were hung on the same day, May 12. Life in the colony went on. But the marching and drilling increased. Virginia was raising a huge militia.

Dunmore, like Gage in Boston, had his back to the wall. He wrote to Lord Dartmouth: ". . . even in the Palace where I live Drums are beating, and Men in uniform dressed with Arms are continually in the Streets, which my Authority is no longer able to prevent."

Now, with everyone incensed about Lexington, the next confrontation could not be far off. It wasn't: Governor Dunmore was about to make his next move.

On May 12, with Patrick Henry and the other leading radicals at a safe distance in Philadelphia and with his family safely back in Williamsburg, Lord Dunmore announced that the House of Burgesses would be reopened on the first Thursday of June. He was convening the legislature, he said, in order to present Lord North's conciliatory proposals, which Lord Dunmore had just received from London. The Crown's plans for introducing martial law into the colonies had been defeated in both Houses of Parliament, and Lord North had switched to a policy of "accommodation." Lord Dunmore hoped it would be well received by the House of Burgesses and would return the colony to its tranquillity.

On May 30, Peyton Randolph surrendered the chairmanship of the Continental Congress to John Hancock in Philadelphia and hastened back to Williamsburg to preside as speaker over the opening of the House of Burgesses. He was led into the capital by a company of cavalry and another of infantry, and was escorted to his home, which was only two psalms and a prayer away from the palace. Church bells rang upon his arrival, and that evening in the Raleigh and the other taverns of the town, volunteer militia raised glasses in toasts.

On June 5, Lord Dunmore opened the new legislative session by reading to the burgesses Lord North's olive branch proposal—the Conciliatory Propositions. North proposed that colonies raise money for their own common defense; the ministry would "forbear to levy any duties, tax or assessment, except only such duties as may be expedient to impose for regulation of commerce."

Lord North's olive branch, however, was grasped in a mailed fist. The Conciliatory Propositions went on to say that Parliament was not renouncing its right to tax the colonies but was merely forbearing the exercise of its right. But that was exactly the point the colonies were refusing to accept.

Unexpectedly, while the House of Burgesses was discussing the North proposal, two young men tried to break into the town magazine and were injured. The cause was a spring-gun trap placed there by the governor.

The House of Burgesses put a militia guard on the magazine. The governor accused the house of usurping the powers of the royal governor without consulting him. The house replied by politely asking the governor to surrender the key to the magazine. With militia troops and cavalry cavorting on the green, the governor believed he was seriously threatened by a hostile house. In the middle of the night—at two A.M.—Dunmore fled once again with his family from Williamsburg to the safety of the HMS *Fowey*.

The House of Burgesses continued to examine Lord North's olive branch and quickly picked it to pieces. Thomas Jefferson was designated to write the official reply: No.

Governor Dunmore now made good on his threat. On the *Fowey*, anchored in Yorktown Harbor, he declared war. He made the incredible move of calling to arms the enslaved blacks and also the indentured servants (many of whom were ex-convicts), offering them emancipation in return. And he called for the Loyalists to raise an army to fight by his side.

Cursing Dunmore's name to the heavens, the planters immediately clamped a nightly curfew on the slaves and put the entire colony on armed alert.

By declaring the slaves free, Dunmore had committed the unforgivable. After this, there was no hope of ever reconciling England and Virginia.

From the sons of some of the loyal planters, Dunmore raised a regiment of Tories that he called the Queen's Own Loyal Virginians. To that he added another regiment composed of slaves to whom he had promised freedom. He named this regiment Lord Dunmore's Ethiopians. Each member wore a legend across his breast: "Liberty to Slaves."

Dunmore scrimmed together an irregular navy that sailed up and down the Chesapeake with little effect. Some 500 Loyalist troops died of disease on the impossibly overcrowded ships.

Richard Henry Lee noted: "If the administration [in London] had searched through the world for a person best fitted to ruin their cause and procure union and success for these colonies, they could not have found a more complete agent than Lord Dunmore."

The governor dispatched an urgent message to a beleaguered General Gage in Boston requesting arms and troops as quickly as possible. With them, Dunmore said, he "would raise such a force from among the Indians[,] Negroes and other persons as would soon reduce the refractory People of this Colony to obedience." He wrote Secretary of State Dartmouth in London that he planned to raise an army to fight the radicals: "I shall not hesitate at reducing their houses to Ashes and spreading devastation wherever I can reach." He would reach, it turned out, not for Williamsburg, but for Norfolk.

* * *

In Norfolk, in September, John Henry Holt, publisher of the *Virginia Gazette or Norfolk Intelligencer* was carrying on a relentless campaign against Lord Dunmore that culminated when he ran a scurrilous genealogy of Lord Dunmore's family. Dunmore, from the deck of the HMS *Fowey,* sent a landing party to arrest Holt. The publisher just managed to escape, but Dunmore's men seized his press and type boxes and two journeymen printers, and with them Dunmore began to put out his own newspaper from the ship.

Holt reimbursed himself partially by stealing two of Dunmore's best horses and set about trying to get the new equipment to start a new paper.

Dunmore now addressed the city of Norfolk.

With a population of 6,000, Norfolk was four times larger than Williamsburg normally and possessed one of the finest harbors on the eastern seaboard. An important component of the population was the Scottish merchants, whose warehouses were filled with considerable wealth. The city was one of the king's few strongholds of Loyalists on the American continent. On New Year's Day, the peripatetic Lord Dunmore sailed the HMS *Fowey* there again. In an inept attempt to strike at local militia and to the inexpressible horror of Loyalists and Crown officers everywhere, he proceeded to burn the city to the ground.

Thomas Jefferson was in his plantation home, Monticello, in the mountains of western Virginia, when news of Lexington reached him.

Up until that very moment, he had been thinking of reconciliation. At first he struggled to hew to that objective against his own emotional reactions. He told himself that the battle was an accident. But as more details arrived in the next days and weeks, the stories of British cruelties fired his hatred for the British. He blamed King George III personally, and thereafter condemned monarchy as an unfit form of government. He also became one of the first outspoken advocates for independence. From the front step of Monticello, his emotional road led him directly to Philadelphia and the writing of the Declaration.

Today, with the long perspective of several centuries, the progression of his thinking from childhood to the presidency seems linear, predictable.

Jefferson's father, Peter, was a strongly independent self-made

Thomas Jefferson

man—surveyor, magistrate, and planter—and a self-styled democrat who married into the conservative, powerful, socially prominent Randolph family when he took Jane Randolph as his wife.

Peter died when Thomas was fourteen, leaving a widow, ten children, and a large plantation. On his majority at twenty-one, Thomas inherited his portion of his father's estate, half of 7,500 acres. His brother, Randolph, got the other half. Peter also left them sixty slaves.

Jefferson, through his own motivation, acquired an excellent education. In 1760, at the age of seventeen, he entered William and Mary College in Williamsburg, and although he was disappointed when he found it ''filled with children, and with low standards of admission,'' he was taught by two men whose influence would remain with him for the rest of his life: George Wythe and William Small, a Scot who taught him math and opened for him the worlds of science and philosophy.

Jefferson also spent a great deal of time watching with fascination the political maneuvering in the House of Burgesses, learning major lessons in practical politics. In time, he would himself be part of that.

He emerged from William and Mary College with much of the intellectual equipment he would need in his public career. He had read deeply in classical literature, history, and natural sciences. He had gained a good knowledge of Latin, Greek, and French; to these he later added Spanish and Italian. The range of his interests was enormous, even remarkable. His restless, ever-searching mind embraced geology, geography, botany, zoology, ethnology, agriculture, medicine and surgery, aeronautics, education, literature and language. He excelled in whatever he touched and was to become a notable architect and a skilled mathematician who used calculus regularly.

Standing six feet two, exceptionally tall for the colonial period, a big-boned, sinewy man with red hair, ruddy features, and hazel-flecked gray eyes, Jefferson towered over most of his associates. Although standoffish among strangers, and a decided introvert, he could be a charming conversationalist, even gregarious when necessary. And while to most he showed a sympathetic engaging personality, he disliked aggressiveness and lacked the charisma that was characteristic of many men of the Revolution. He was, in fact, a loner who much preferred his own company, his books, and his violin. A gifted musician, he often played three hours a day.

He was widely admired for his excellent horsemanship, and early in life took a passionate interest in thoroughbred horses, which he described as "a necessary luxury." In an age of unbridled, ruinous gambling, he never played cards—never allowed cards at Monticello, never indulged in gambling in any form, never used tobacco, and "was never party to a personal quarrel."

Early in his life, he had difficulties accepting much of what he learned about the social structure in England. And as an adult, his revulsion with the injustices and corruption of English society was well known.

In the Williamsburg office of George Wythe, Jefferson read for the law, often fifteen hours a day. For a while he even abandoned his beloved violin. Wythe presented Jefferson with a powerful intellectual tool when he taught him how to combine lucid principle and a point-by-point application of it, a methodology that left its stamp on everything Jefferson wrote and thought thereafter.

When Jefferson opened his own law offices, at the age of twenty-four, success came quickly with a roster of distinguished clients. In the seven years that he practiced law—before abandoning it for politics—he earned the exceptional income of £300 a year. In 1774, at thirty-one, he left his practice with a poor opinion of lawyers; "whose trade it is to question everything, yield nothing, and talk by the hour."

Jefferson revealed another side of his genius in February 1770, when he began excavation for Monticello, his home, his lifelong passion, and his monument. Before Monticello was finished, his family house, Shadwell, burned to the ground. The fire consumed everything, including his cherished collection of books. The only item to survive was his violin—which he described as "not a very good one."

On New Year's Day 1772, he married Martha Wayles Skelton, a wealthy twenty-three-year-old widow, the daughter of a rich Williamsburg lawyer, and brought her to Monticello, which was still under construction. She was described as "very handsome and accomplished." Her harpsichord had married his violin. It was a notably happy union until 1782, when she died at the age of thirty-three during childbirth. Of their six children only two—both daughters—survived infancy, and although Jefferson was a devoted and affectionate father and grandfather, he never remarried. His marriage to Martha had added over 11,000 acres to his holdings, which put him well on his way to becoming—like his friend George Washington—one of the largest landowners in Virginia.

Jefferson entered the House of Burgesses in 1769, and he soon aligned himself with other radicals such as Patrick Henry. By 1773 he had become a member of the radical Virginia Committee of Correspondence.

By turns a dedicated scholar and a skillful political manipulator, he worked best in the one-on-one give-and-take of small committee meetings behind closed doors. There his influence was pervasive. People who worked with him discovered that in an age when skillful oratory was esteemed, even essential, Jefferson did not talk spontaneously and was not fond of public speaking.

He said he hated "the morbid rage of debate . . . believing men were never convinced by argument, but only by reflection, through reading and unprovocative conversation."

Even though he was an important member of many committees and deliberative bodies, in all his time in the House of Burgesses, he never made a speech. Instead, he much preferred to express himself in writing, putting his thoughts on paper, revising them, polishing them, then revising them again. A slow process it was, but as a writer he excelled; he was probably the most inspiring writer of the entire Revolution. While many of the speeches of his associates were soon forgotten, most of his writings have survived intact as an essential element in the intellectual founding and development of the new nation.

His great writing gift was first revealed in 1774, when he published "A Summary View of the Rights of America." It was written originally as instructions to the Virginia Convention, which was to meet to discuss intercolonial union. The work of the Virginia Convention was ultimately to lead to the Continental Congress.

In the "Summary View," he expressed his chagrin at the control that the upper classes maintained over the entire population of England. And when he noted that on that island of six or eight million, the entire House of Commons was elected by only some 160,000 patrician voters, his sense of injustice was deeply offended—not only for the English but more particularly for the Americans: "Can any one reason be assigned, why one hundred and sixty thousand electors in the island of Great Britain, should give law to four million in the States of America, every individual of whom is equal to every individual of them in virtue, in understanding, and in bodily strength? Were this to be admitted, instead of being a free people, as we have hitherto supposed, and mean to continue ourselves, we should suddenly be found the slaves, not of one, but of one hundred and sixty thousand tyrants."

In his "Summary View," Jefferson's ideas were, as usual, far in advance of those of his fellow politicians and, for many conservative planters, far too radical. Jefferson denied emphatically the right of Parliament to govern the colonies and enumerated all the errors of the king's colonial policies. Consequently, his "Summary View" was rejected as too radical to guide the convention. However, there were many who were strongly impressed by it. Slightly modified, probably by Edmund Burke, the "Summary View" made a great impression in England and put Jefferson in the forefront of the American Revolution.

Among the American radicals, Jefferson was now recognized as a powerful and inspiring writer. Due to the stature of the "Summary View," Jefferson was later to be invited to compose the Declaration of Independence—sections of which were an outgrowth of the "Summary View."

Had he continued to harbor even the faintest shadow of doubt about the rightness of breaking with the British monarchy, his visit to France emphatically erased it. Paris was his first real opportunity to study monarchy up close, and it confirmed his hatred for arbitrary power. He found life in France under Louis XVI unspeakably wretched. He wrote to Washington:

> I was much an enemy to monarchy before I came to Europe. I am ten thousand times more so since I have seen what they are. There is scarcely an evil known in these countries which may not be traced to their king as its source, nor a good which is not derived from the small fibres of republicanism existing among them. I can further say with safety there is not a crowned head in Europe whose talents or merit would entitle him to be elected a vestryman by the people of any parish in America.
>
> Of twenty millions of people supposed to be in France . . . nineteen are more wretched, more accursed in every circumstance of human existence, than the most conspicuously wretched individual of the whole United States.

The other leading figures of the Revolution—John Adams, Hamilton, Washington, Patrick Henry, even Samuel Adams—brilliant leaders though they were, were primarily political revolutionists, determined to replace one political system with another. Jefferson was in revolt against any tyranny "over the minds of men," seeking not only to free the colonies from their ties to the British Crown but also to make fundamental—and democratic—changes in American society itself.

Only Jefferson was a revolutionist of the human spirit. Only Jefferson could have written "I have sworn upon the altar of God eternal hostility against every form of tyranny over the mind of man."

Jefferson's horse-drawn surry—known as a phaeton—familiar along many of the rural roads of Virginia, traveled with two spare horses hitched to the back as he made the ten-day journey to Philadelphia from Williamsburg to replace Peyton Randolph.

He carried with him a document that would help add significantly to his recognition as one of the intellectual leaders of the American

Revolution and also bring the Revolution several big steps closer—the reply to Lord North's proposal.

In Charleston, South Carolina, on the same night that Governor Dunmore removed the gunpowder from the Williamsburg magazine, April 21, a volunteer group of South Carolina radicals took both arms and ammunition from the town armory.

The ripple effect from Lexington had continued south with unsettling results. A short time after the news arrived in New Bern, North Carolina, Governor Josiah Martin, feeling control slipping from his hands and confronted by an increasingly truculent population, uttered the same threat as Governor Dunmore to free the slaves and arm them. The response was so fearsome that Martin fled from the governor's residence in New Bern and prepared for a siege in Fort Johnson, at Cape Fear.

Unwittingly, he had terminated royal government in North Carolina. For by July 17 the governor realized he could not defend himself at Fort Johnston, which was badly deteriorated and, faced with an armed assault, he again emulated Governor Dunmore and withdrew to the safety of the sloop-of-war HMS *Cruiser,* taking the fort's guns with him.

That made two royal governors since the battle at Lexington floating offshore from their respective colonies, out of their capitals, and out of power, never to return.

The news of Lexington reached Fredericksburg, Virginia, and the ears of John Harrower on April 28. Harrower, a forty-year-old bankrupted Scottish merchant and businessman, and now an indentured servant employed as a tutor, heard it with a sinking heart. A recent (1774) arrival, he was increasingly enthusiastic about becoming an American, gaining his freedom, and bringing his family from Scotland. But if Lexington meant war, it also meant no passage for his wife and young children, now waiting for ship money in Lerwick, on the Shetland Islands of Scotland.

Harrower was one of a legion—a vast army, in fact—of indentured servants in the colonies. For those who could not afford the cost of passage, one way of immigrating to the great opportunities in the New World was by signing a contract of indenture for a set period of time—often four years, as in Harrower's case. The conditions they labored under were frequently so harsh that indenturing was often called white servitude.

For colonists confronted with a chronic shortage of labor in America, buying an indenture was a popular way of circumventing this shortage.

Harrower, after bankruptcy caused by the severe economic depression in Scotland and England in 1773, left his wife and children and spent months wandering in Scotland and then in England, trying to find work—any sort would have served. He was one of a flood tide of Scots who all through the 1770s were abandoning their homeland and the unending lack of employment there. At last, having wandered from seaport to seaport, he turned toward London and traveled the eighty miles from Portsmouth on foot; he found nothing there but starving workers. Feeling he had no other choice, he signed papers to become an indentured servant in America.

Passage on the ship *Planter* from England to Fredericksburg, Virginia, took almost seventeen weeks. Harrower helped nurse the sick on board—both passengers and crew—and even filled in for the chief mate by keeping the ship's journal and log. The captain had been so impressed with Harrower that he praised the man highly to William Anderson, a Fredericksburg merchant who sold servants' indentures.

Fredericksburg, a thriving port on the Rappahannock River, was an important trade center surrounded by a number of fertile plantations; it was also in the center of the rapidly growing Piedmont region. In addition, it was the gathering place for wealthy planters and their families. This meant that Fredericksburg offered many amenities, including frequent horse racing, theatrical performances, balls, concerts, and socializing. George Washington was known to have spent many a convivial evening in the town's taverns.

On Fair Day came Colonel William Daingerfield, who was in need of a tutor. Anderson recommended Harrower. It was a fortunate meeting for the new immigrant. The wealthy Colonel Daingerfield was prominent both politically and socially in northern Virginia, had been educated in England, and was an openhanded, generous man. Unlike the typical indentured servant, Harrower found himself eating at the family table, socializing with Daingerfield's visitors, and even attending church with the family.

Belvidera, the Daingerfield estate, was Harrower's first close-up experience with real wealth. And he had come at a time when Daingerfield, like other planters, was coping with the soil exhaustion tobacco caused. He planted wheat and corn in its stead—a process that fascinated the Scottish tutor.

Harrower was awed by the whole New World he had entered and wrote glowing letters to his wife. "I never lived a genteel regulare life untill now," he wrote. He assured her he was working to bring her and their "sweet infants" to America as soon as possible. "I yet hope please God," he wrote his wife, "if I am spared, some time to make you a Virginian Lady among the woodes of America which is by far more pleasant than the roaring of the raging seas round abo't Zetland."

Harrower was fortunate to have gained employment as a tutor because there was a growing prejudice among the planters against Scottish schoolmasters, who implanted their Scottish burr in plantation offspring. This burr, complained one critic, "they can never wear off."

Also, Harrower had the option of teaching other children for a fee, which hastened the day when he would have enough saved to send for his family. Evidently, Harrower must have had some success in modifying his Scottish accent, for he soon earned a good reputation as a teacher and gathered more children into his classroom. In time, he took on the challenge of educating John Edge, a deaf-mute boy, and also began to instruct slaves in the catechism.

In his diary and in his letters to wife, he recounted with great interest the momentous events leading to the Revolution, especially the events in Boston, which he presented with considerable accuracy and a notable bias toward the colonists. Since relations between the scornful English and the poverty-stricken Scots was very poor, it is probable that Harrower brought to America the same animosity toward England most other Scots bore.

Even as early as December 1774, Harrower was writing to his wife about booking passage to America. "Take your Passage to Leith, from thence go to Glasgow and from that to Greenock where you will ship for this Country." In April the battle at Lexington dismayed him: "Friday 28th This day by an express from Boston we are informed of an engagement betwixt the British troops and the Bostonians, in which the former were repuls'd with loss, but no particulars yet."

With the port of Boston already closed and others likely to be closed soon, there would be no family crossing until the hostilities ended—a long way away probably, unless he found an alternate route.

Harrower managed to save £70 toward the passage of his wife and children from Scotland to Virginia. But tragically, in 1777 at the age of forty-four, he died from an unexplained disease.

His kind patron, Colonel Daingerfield, also died a tragic death. After the Revolution, during a severe economic downturn, the colonel suffered such heavy losses on his plantations that he faced complete ruin and a life of poverty. Deeply depressed and unable to accept that fate, in January 1783 he cut his own throat.

The brigantine *Industry,* bearing news of Lexington, reached Charleston, South Carolina, on May 8 and encountered a strangely muted reaction on the docks of the city. In the Carolinas and Georgia there was a greater preponderance of Tories than Whigs. Still, they numbered themselves among the most defiant in all the colonies of Parliament's efforts to tax them because, as the greatest shippers of raw materials and importers of finished goods, they felt they would have borne the greatest weight of British taxes.

In fact, during the Stamp Act crisis, Christopher Gadsden, one of the province's wealthiest men, mobilized his social club and led them in riot. When the mob swept into the home of stamp official Henry Laurens to destroy the tax stamps, despite soot and disguises Laurens recognized a number of well-to-do gentlemen of the town. They found no stamps, but they got into Laurens's booze cellar and merrily tottered off to attack Fort Johnston. Chief Justice Shinner wisely greeted other rioters at the door with punch bowls brimming. They all toasted damnation to the Stamp Act and went home soused, their mission forgotten.

Gadsden opined that by passing the Stamp Act the Grenville Ministry "must have thought us Americans all a parcel of apes and very tame apes, too." During the tea crisis, Charleston defied England with a good more decorum than Boston. The tea was seized by customs and stored out of harm's way—the exact opposite to the situation in Boston. No one in Charleston ever paid a penny in tea tax. Yet the southern colonies were temperamentally no more ready to make a break with England than their northern counterparts. Instead most Carolinians believed that Sam Adams was trying to drag them into a revolution they didn't want.

Gadsden was a particularly pugnacious opponent of Parliament and was often called the South Carolina Sam Adams. He had already attended the First Continental Congress and, previous to that, the Stamp Act Congress in New York City, where he actively fought against putting any obsequious phrases in the documents the congress created—and won. The bluntness should have been, but wasn't, a warning to Parliament.

Although the image of wealthy, conservative property owners taking the part of rioting revolutionaries may seem to have had comic opera overtones, there was nothing comic about the participation in the Revolution of the Carolinas, where some of the sharpest fighting took place.

Shortly after the news from Lexington arrived, Christopher Gadsden hurried northward to the Second Continental Congress, where he joined with George Wythe, the two Lees of Virginia, and John Adams against any consideration of reconciliation with Britain. His was a loud and an important voice.

Preparing for War

While the news of the battle spread slowly southward, the militia in Massachusetts was busy.

In Watertown, bookseller Henry Knox, wearing a handsomely tailored uniform and highly polished boots, and now dubbed "Colonel" by his troops, could look along the fortifications that stretched for ten miles around Boston, from the Mystic River, sweeping south through Cambridge, along the Charles River, curving around to Roxbury, and curving around even more to Dorchester Heights. In that whole sweep, Knox had the strongest emplacements. But not one cannon. And no gunpowder. It was now April 30, and Boston was still quiet. Knox was wondering what General Gage was up to.

Disease appeared with shocking speed around Boston. By April 28, camp fever had spread outside the city. By the time the news of Lexington had reached Williamsburg, camp fever had already broken out among New England's militia and had reached the general population. And by late August it would gather thousands to the grave.

The sick lay moaning in delirium along the roadside, under bushes, in barns and stables and sheds, dying in the midst of the filth of dysentery.

Before the epidemic was stopped by the cold weather, John Adams hastened to his sick soldier brother, Elihu, in his Roxbury camp, but he reached him too late. He carried the body home for a burial. Piling grief on grief, Abigail Adams tried to nurse back to health her cherished mother in the old Weymouth Parsonage, only to lose her to dysentery.

To John Adams's great terror, back in Philadelphia, he heard that Abigail and four of their children had come down with the deadly camp fever. His joy was inexpressible when all five survived.

Inside Boston, General Thomas Gage was still strangely inactive. He was doing nothing while the Continental army organized itself and grew stronger. By May 3 he had missed opportunity after opportunity to confront the militia and to take the initiative. And he had gained nothing by waiting, for General George Washington would soon be on his way to take command.

Because of his vacillating, remaining immobile inside the city day after day, his troops now called him "The Old Woman."

The British army sentry on Boston Neck and the American sentry from Roxbury were described by Gage as being a musket shot apart. Something had to happen soon.

Isaiah Thomas, editor and publisher of the *Massachusetts Spy,* had set the type for his first edition from his new base in Worcester and now stood looking at his first proof sheet. He was about to publish the greatest story in his journalistic career. Rebelliousness shook every word of the story. It was May 3.

The same rebelliousness had been a dominating component of his complex character since childhood. His desperate widowed mother had apprenticed him at the age of six to the printer Zechariah Fowle, indenturing him until his twenty-first birthday. He got his only formal education right at the typecase, his child's fingers picking type for such textbooks as the *New England Primer* and *Pater's New Book of Knowledge.* But as he grew, he chafed; he wanted to learn more about the printing trade, he wanted to finish his apprenticeship under a world-class printer in England, he wanted his freedom.

When he was sixteen he ran away to Nova Scotia, where he was given free reign of the *Halifax Gazette* by its complaisant owner. Thomas quickly trumpeted the alarm about Parliament's newly imposed Stamp Tax. When his invective reached treasonable levels, the owner ejected him. Unable to get passage to England, Thomas returned to Fowle, where he carried on his campaign. Although still a teenager, he had become a fire-breathing American patriot with a quickly developing talent for inflammatory propaganda. The radicals in Boston, especially Sam Adams and John Hancock, took note of him.

But restless still, tall, strong, and ambitious, he was soon off again, this time south. In Wilmington, North Carolina, where he tried to set up a newspaper, a lady who owned a tavern talked of a partnership. When the newspaper deal fell through, Thomas's prolonged farewell to the lady was so heedless that he missed his ship's sailing, and several burly rowers had to race after his departing ship to get him on board.

In Charleston, South Carolina, he put the finishing touches on his publishing apprenticeship by working several years for a very efficient Tory newspaper owner, during which time he must have effectively concealed his extreme radical inclinations.

On Christmas Day 1769, he married Mary Dill, a girl from Bermuda. She was nineteen; he was twenty. In January, with his new wife at his side and harboring a burning desire to produce the finest newspaper in the colonies, Thomas returned to Boston. He was ready to settle down in his own shop. His wife was hardly ready to settle down in her marriage.

After fathering two children, Thomas made some shocking discoveries about his wife's past, and he sued Mary Dill for divorce in no uncertain terms. He told the court that his wife had given birth while a teenager to a "bastard sone" and that "she had been prostituted to the purposes of more than one."

Cuckold's horns or no, in early 1770, Isaiah Thomas and Fowle started the *Massachusetts Spy*. It quickly succeeded, and by November Thomas had bought Fowle out with money supplied by John Hancock.

His *Spy* was a fine product, with handsome typography in a large-sheet, four-page format. The text was set in a gracious, readable typeface, but the words were all cannons aimed right at Britain and her royal officers—particularly the royal governors, who came and went with dismal regularity.

Though barely out of his teens, Thomas showed an astute business head combined with a dizzying skill at abandoned invective. He brilliantly used the propagandist's craft of seamlessly insinuating opinion into the news, and his defiant patriotism equaled that of

Gill's and Edes's *Gazette.* His circulation climbed steadily to become the largest in New England. Among the wildly incendiary newspapers of the colonies, the *Massachusetts Spy* was the most incendiary of them all.

Thomas narrowly escaped prison and the end of his journal when fellow radicals helped him flee from Boston with his type boxes and press just before the Lexington and Concord battle. He moved everything to Worcester, forty miles inland, well beyond British authority.

Now, with his new base, free from all partners and business associates—and away from the potential interference of the British army (Thomas had not forgotten the threat of the entire British 47th Regiment, including the marching band, to tar and feather him)—he was eager to carry on.

Across the top of his front page, above the masthead:

Massachusetts Spy
Or, American ORACLE of Liberty

Thomas sniped a shocking headline:

Americans!—Liberty or Death!—Join or Die!

Thomas then proceeded to recount the events of the battle, making sure the reader understood that the British soldiers fired first:

> Americans! Forever bear in mind the BATTLE at LEXINGTON!—where British Troops, unmolested and unprovoked, wantonly, and in a most inhuman manner fired upon and killed a number of our countrymen, then robbed them of their provisions, ransacked, plundered and burnt their houses! Nor could the tears of defenceless women, some of whom were in the pains of childbirth, the cries of helpless babes, nor the prayers of old age, confined to beds of sickness, appease their thirst for blood! Or divert them from their DESIGN of MURDER and ROBBERY.
>
> It is to be noticed they fired upon our people as they were dispersing, agreable to their command, and that we did not even return the fire [actually, they did] . . .

He did not hesitate to color the events with editorial asides for the reader's better comprehension:

> . . . others of them, to their eternal disgrace be it spoken, were robbing and setting houses on fire and discharging their cannon at the meeting house.

Thomas's account of the battle at Lexington and Concord would be the first that many colonists would see, and it would be widely copied by other newspapers, complete with the outrage and fulmination against Britain. In the *Spy,* weeks, even months, ahead of public opinion, there would be no more talk of reconciliation with Britain. Lexington had closed that door.

Two days later, on May 5, the Massachusetts Provincial Congress, during its meeting in Concord, resolved that General Gage had "utterly disqualified himself to serve this colony as a governor and that no obedience be paid . . . to his writs." He was accused of being "an unnatural and inveterate enemy to this country."

On the same day, May 5, in Philadelphia Benjamin Franklin arrived at the dock from London. After eleven years in London working to keep the colonies and Britain in a working harmony, Ben Franklin walked down the gangplank with his grandson, Temple, at Philadelphia's Market Street Wharf in the mild night air and onto the docks of Philadelphia, where he received a tumultuous welcome from the throngs who had come to see the famous doctor.

And there he received solemnly the news of Lexington and Concord.

The crowd followed him up the street to the home he had commissioned to be built but had never seen in its finished state. For the rest of his life, he was to share the house with his daughter, Sally Bache, her husband, and his seven grandchildren. He had not seen Sally in eleven years. And now he met her seven children for the first time.

The next morning, Franklin conferred with his political allies and was selected as a Pennsylvania delegate to the Continental Congress. He was an enormously valuable asset to the congress because of his firsthand knowledge of London, Parliament and the royal cabinet.

Franklin then turned to other unfinished political business. This would prove personally more painful.

Benjamin Franklin had returned to a strange new world. His wife and many of his old friends were dead. Other friends had become enemies. The city had changed, values had changed, and a younger generation had taken over. He was widely admired as an elder statesman. But his work—his public service—was far from over.

One word described the stature of Benjamin Franklin when he stepped off the ship's gangway: *colossus.* Save for Thomas Jefferson, none of his contemporaries had done so many things so well.

An exceptionally gifted writer, a publisher of great influence, a scientist, a philosopher, an indefatigable inventor, a philanthropist, a politician, a statesman, and a diplomat, he had achieved a list of accomplishments reaching back into his youth that was truly astonishing.

He was the tenth son in the family of seventeen of a Boston candle maker. After a five-year printing apprenticeship under his half brother, at seventeen he ran away to Philadelphia. In time he taught himself algebra and geometry, navigation, grammar, logic, writing, as well as both natural and physical sciences. He patiently learned German, French, Latin and Spanish. Almost completely self-taught, he became one of the best educated men of his time, one of the leading scientists in the world.

Owner of his own print shop at age twenty-four, he became publisher of the *Pennsylvania Gazette,* a newspaper of great influence and with the largest circulation of any paper at that time. He was also writer-publisher of *Poor Richard's Almanac,* which had annual sales second only to the Bible. In a relatively short time and while still in his twenties, he had become the most successful printer in the colonies.

In his mid-forties, he was so wealthy that he retired from his printing business and thereafter devoted himself mainly to political matters, spending all told some eighteen years in London representing Pennsylvania and other colonies during the fractious times of George III's early reign.

He was also among the most celebrated scientists of his time, one of the first men in the world to work with electricity. During a thunderstorm, he demonstrated with just a kite and a key, that lightning is electricity—then, to protect against it, he invented the lightning rod. He was the first scientist to study the movement of the Gulf Stream. He invented daylight saving time so that people would not "live much by candlelight and sleep by sunshine." Among his multitude of other inventions were bifocal glasses and the first efficient wood-burning stove, which for the first time made household comfort in the dead of winter a practical reality, rescuing people from distracting cold with a fraction of the wood required by the highly inefficient and dangerous fireplace.

To make sure that his inventions were of unencumbered use to others, he did not patent any of them.

Not all his experiments were solemn or successful. Trying to determine the lethality of electricity, he attempted to use lightning to execute a turkey. When instead he himself was briefly knocked unconscious by the lightning, he said, "I tried to kill a turkey and nearly killed a goose."

The scope of his public service still awes. He established Pennsylvania's first college, which later became the University of Pennsylvania. He established the city's first public hospital which became today's Pennsylvania Hospital. As deputy postmaster general of the colonies he greatly improved and modernized the entire postal service. He organized Philadelphia's first fire company, desperately needed in a city with many wooden houses, reformed the police department, and led the program to pave, clean, and light Philadelphia's streets. He founded the world's first subscription library, founded the American Philosophical Society and became the first president of the Pennsylvania Society, for Promoting the Abolition of Slavery. He served as clerk of the Pennsylvania legislature and is remembered today also as a signer of the Declaration of Independence, which he helped Thomas Jefferson to edit, and as a participant in the creation of the United States Constitution.

When Franklin first sailed to England, and for many years thereafter, he believed that in time the colonies would be the dominant element in the British Empire, the capital would be in America, and England would be subordinate.

The turning point in Franklin's attitude toward Britain had occurred during the notorious Wedderburn encounter. This was brought about when Sam Adams violated Franklin's confidential report and so exposed him to ruin.

In June 1773, when relations between the Crown and its colonies were reaching the boiling point of the Boston Tea Party, Franklin was still working diligently for reconciliation. A friend and confidant in London delivered to his hands a packet of very revealing personal letters, most of them written by the governor of Massachusetts, Thomas Hutchinson, to the late Thomas Whately, one of the authors of the Stamp Act.

On the surface, there seemed to be nothing very newsworthy in the letters. In fact, some of them were six years old. The governor expressed his thoughts on the political conditions in Massachusetts, revealing his distrust in the ability of the common people to govern themselves and his fear that they were prey to demigogues, and suggesting that the colonists had too many liberties. London, he felt, should take a firmer hand in Massachusetts. But Hutchinson had never uttered such thoughts in public.

To Franklin, who was trying to resolve transatlantic differences, the letters seemed to reveal a primary source of the hostile opinions held by the cabinet in London. These letters might enlighten the Boston radicals and encourage them to modify their stiff-necked stance. So he sent them on to Massachusetts House Speaker Cushing with several very specific restrictions. The letters were sent in the strictest confidence and were for information only. Only the top few Boston radicals would be allowed to read them. Furthermore, Franklin emphasized, there could be no publication, and under no circumstances were any copies to be made. The originals were to be returned to Franklin. Failure to follow these stipulations could lead to Franklin's ruin.

Samuel Adams, who disliked Franklin and whose scruples rarely stood in the way of his objectives, violated Franklin's instructions and made copies of the letters, then announced the mysterious appearance of the copies. When he had the letters read aloud in front of the entire assembly, Franklin's ruin was assured. Adams then went on a ranting newspaper attack against Hutchinson, whom he hated passionately, publishing extracts of the letters, "wrenching and spraining the text," taking words out of context, and finding thoughts and implications that were not present in the original letters. Adams's journalistic vilification of Hutchinson went on for weeks.

Next, based on the letters, Samuel Adams had the Massachusetts Assembly send a petition to London for the removal of Hutchinson as governor of the colony.

Whately's son was furious, accused Sir John Temple, a staunch Whig, of stealing the letters from his father's estate, and fought an inept duel over the matter. To save Temple from further harm, Franklin revealed that he himself had sent the letters to Boston. That was in December. In January 1775 the Privy Council in London took up the petition to remove Hutchinson and announced a hearing in the Cockpit Room at Whitehall. The real intent of the meeting was not to discuss the petition but to attack Franklin for revealing the letters. The sixty-eight-year-old statesman was summoned to appear before the thirty-six members of the Privy Council plus a number of observers, including Lord North, Lieutenant General Thomas Gage, General Sir Jeffrey Amherst, Edmund Burke, and the physicist Dr. Joseph Priestly, a radical and a longtime friend of Franklin.

The meeting, held in the Cockpit Room in January 1774, was conducted by the solicitor general, Alexander Wedderburn. This was an unfortunate turn of events for Franklin. Wedderburn, who had gone to great lengths to eliminate his Scottish accent, was one of the deadliest debaters in Parliament. Vicious, ambitious, totally without principle, and despised by the king, he was quite capable of taking either side in any debate.

Wedderburn stood up in the packed chamber and for one hour pilloried Franklin with a savage and outrageous personal attack. He called Franklin a wily New Englander, a thief, and much more: "Men . . . will hide their papers from him, and lock up their escritoires." During the whole attack, Franklin stood, head up, in silence while many of the observers openly laughed at him. But a few were shocked. Burke condemned Wedderburn's performance as "beyond all bounds and measure." Lord Shelburne labeled it "a scurrilous invective." The *London Packet* of February 2 termed the attack "the choicest flowers of Billingsgate." But other papers attacked Franklin as a viper, a scheming provincial, an old dotard, a skunk, a polecat. Dr. Johnson called Ben Franklin, "the Master of Mischief."

His disgrace was completed the next day when he was removed as deputy postmaster-general of the American colonies, a position worth £300 pounds a year. Completely humiliated, Franklin was at the lowest point in his career in England.

After fighting for years to preserve the relationship with England, Franklin finally saw the impossibility of continued union. He wrote to Galloway: "When I consider the extreme corruption prevalent among all orders of men in this rotten state [England] and the glorious public virtue so predominant in our rising country, I cannot but apprehend more mischief than benefit from a closer union. Such a union would be like coupling and binding together the dead and the living."

The British government, he now believed, wanted "to provoke the North American people into open rebellion." Independence was the only colonial course left. From this point on, Franklin, the most dangerous man in America according to some Britons, would be a major factor in the wresting of the colonies from George III.

Years later, in 1783, Franklin was to have the last word when, after inducing France to join forces with the Americans, he went to sign the Paris peace treaty that severed the thirteen colonies from Britain. He wore the same suit he had worn during Wedderburn's attack. He had saved it all those years just for this occasion.

By March 1775, just weeks before Lexington, Franklin had decided it was time to get out of England. Hostility had increased so much within the British government that he feared he might be arrested and imprisoned. Feeling his long years in London "had served little purpose," he left England on March 21 and returned home to a revolution.

General John Thomas, watching Boston from Roxbury, had a strong expectation that General Gage would sally forth from Boston, drive down the Boston Neck, and attack Roxbury, which was protected by only 700 militia. So on May 9, to convey the impression of a much larger force, General Thomas ordered his 700 men to march around and around on a high hill fully visible from Boston. There was no attack from Boston. General Gage remained immobile.

In the Pennsylvania State House in Philadelphia, the Second Continental Congress was called to order while the two ships carrying the news of Lexington to London were still on the high seas, probably as much as three weeks away from Southampton. It was May 10, and already Philadelphia was hot and humid.

During the first assembly of congressional delegates, John Adams was near tears when he greeted the delegate from Georgia, Dr. Lyman Hall, who had ridden six consecutive weeks on horseback, more than 800 miles, to attend.

Dining in taverns throughout the city, with Lexington the principle subject of talk, the delegates were becoming reacquainted, feeling each other out for changes in viewpoint caused by recent events. Sam Adams believed he was in for a tough fight. Despite the events in Lexington, he heard little sentiment for a break with Britain and much talk of conciliation. John Adams said he was amazed.

When, after a steady march, Benedict Arnold had arrived in Charlestown with his militia, he went before the Massachusetts Committee of Safety and proposed that a contingent of 400 men under his command be sent to capture Fort Ticonderoga, a key military location that dominated the entire Hudson River Valley. The Committee of Safety commissioned him a colonel in the new American army and authorized him to recruit 400 men to implement his plan.

And so it came to pass that on the same day the Second Continental Congress opened its session in Philadelphia, British army captain William Delaplace, commandant of Fort Ticonderoga, learned of the battle of Lexington in a shocking manner.

He was rudely awakened in his bunk by Ethan Allen shouting "Come out, you rat!" found a musket pressed to his nose, and soon discovered that his fort had been attacked and overrun by Benedict Arnold, Ethan Allen, and 150 Green Mountain boys. The Americans were exultant over the large number of cannon and barrels of gunpowder they had captured.

Later, Henry Knox would drag those cannon to Dorchester Heights, overlooking Boston Harbor, to force the British evacuation of that city.

Joseph Galloway, not well and fearful of the radicals who had come to Philadelphia for the congress, sent his carriage to bring Franklin from Philadelphia to his country seat in Trevose in Bucks County.

When they met and greeted each other—it had been over a decade since their last meeting—Galloway looked much older and sickly. And the meeting, which he had looked forward to with such hope, soon deeply disappointed him. His old comrade in arms—with whom he had fought many political battles—was in the exact opposite political corner from him. Galloway had become even more of a royalist, more outraged by the mobbing he had gotten at the hands of that great leveler and street politician, Sam Adams. Franklin, also a victim of Adams but equally outraged by the terrible ordeal that Wedderburn had put him through, had become completely radicalized. For him there could be no reconciliation, should be no reconciliation, with the Crown. Franklin's recent letters had warned Galloway how completely he had turned his back on England, but Galloway had remained hopeful to the end.

It was soon apparent they were in a hopeless impasse. Galloway wanted to convince Franklin and Franklin wanted to convince Galloway and neither would yield.

Trying one more time, Franklin met with Galloway in Trevose in June. This time Franklin brought his London journals with him, read from them, gave Galloway chapter and verse on the corrupt and venal society of London, hoping to open Galloway's eyes to the overriding fact that the two countries were doomed to separate. He failed.

The two old friends could only look sadly at each other.

They even tried a third time a few weeks later. This time Galloway invited Benjamin Franklin's son, William, with whom he had formed a warm personal association, almost a father-and-son alliance. William Franklin was royal governor of New Jersey and as stout a Loyalist as Galloway. In fact, he had helped Galloway write his Plan of Union prior to the First Congress.

Now Franklin was confronted by the two of them—old friend and son—and not budging, failed to budge them. Franklin's relationship with his son was beginning to deteriorate. The old man believed that William had fallen under the influence of his strongly Tory wife. He would become estranged from William in time.

Joseph Galloway resigned from his delegate's seat in the congress and withdrew from public life in Pennsylvania. He would soon associate himself with the British army.

In Massachusetts, the militia continued to try to goad General Gage into some kind of action. On May 13, under General Putnam, an army of 2,200 men, extending a mile and a half, marched down Bunker Hill and Breed's Hill. They entered Charlestown by the fish market, near the ferry, in full range of the guns in Boston and on the HMS *Somerset,* so close it was said they could see the sailors clearly, some holding burning fuses to fire the loaded cannon. No shots were fired, and the Americans marched back to Cambridge. Yet, despite the personal reluctance of General Gage, the will to fight on both sides was growing stronger.

The military situation in Massachusetts had gotten quite unmanageable, and on May 15, the Provincial Congress of Massachusetts sitting in Concord, sent an express rider—Dr. Church—to Philadelphia. The message Dr. Church carried urged the Continental Congress to set up a continental government to take charge of the militia surrounding Boston.

Two days later, on May 17, as if to punctuate the dangerous state of affairs, a great fire broke out in Boston, lofting smoke high above the city and visible for miles. It began in a military barrack on Treat's

Wharf and burned twenty-seven stores, one shop, and four sheds. General Gage had recently appointed several new captains to command the engine companies. This act offended the engine men who were summoned to put out the fire: "Hence, the engines were badly served." The cause of the fire was not known.

On May 18, with great dismay and shock, the congress heard about the capture of Fort Ticonderoga. An order was issued to make a careful inventory of the fort and its contents against the day when it would be returned to Britain.

The congress still fought against the idea of independence. John Adams, presenting a comprehensive military plan for fighting the British, which was certainly essential, "saw horror and detestation" on many faces.

The pressure was building on everyone. The Reverend William Smith gathered with five other Anglican clergymen in Philadelphia to write a letter to the bishop of London. They had an immediate and alarming problem. Delegates to the Second Continental Congress, meeting a few blocks away in the Pennsylvania State House, had just ordained Thursday, July 20, as a day of fasting and prayer.

To the six ministers this meant that on that day they would have to call down blessings on the efforts of the congress to assert its rights against the unwarranted intrusions of Parliament into colonial affairs. There could be no evasion on that day. As official members of the Church of England they would be expected to express sentiments that could be construed by their fellow Anglican ministers in England as disloyal. Disloyalty was a hanging offense in London.

They were keenly aware that their letter would be read by their bishop, no doubt already shocked and concerned by the news from Lexington.

"We sit down under deep affliction of mind," it began, "to address your Lordship on a subject in which the very existence of our Church in America seems to be interested. . . . Our complying [with the observance of the day of fasting and prayer] may perhaps be interpreted to our disadvantage in the parent state."

Not to reject parliamentary taxation and extensions of power into colonial affairs would be seen by the radicals as disloyal to America. Did the bishop of London expect them to utter from their pulpits resounding condemnation of the proceedings of the congress? To demand loyalty to the king and his church? Reverend Smith could quote an apt phrase from his previous sermon. Such behavior "would be destructive to ourselves, as well as to the Church of which we are ministers." The word, *destructive* could be left to resonate with multiple meanings for the bishop's contemplation.

They knew that on Thursday, July 20, into their churches would come fellow Anglicans such as George Washington expecting to hear the politically right words come tumbling from the pulpits. So, the letter explained, the six ministers had, after long and agonizing discussions, decided to express at least nominal sympathy with the radical agenda. As they affixed their names to the document, they knew the missive would stun their colleagues in England. And they regarded their decision with great dismay.

But even as the letter was set on its weeks-long journey across the Atlantic Ocean, they were aware of an even larger, more frightening, issue. Of all the disloyal words being bruited about the city by the delegates, the most frightening was the word *independency,* which had been whispered in the taverns and even on the church steps of the city. If liberty did come, were they then to set up a new church, independent of Britain? None of the six were ready for that.

By mid-May, the news had traveled over the mountains into the Ohio Valley and had reached, among others, a party of hunters in Kentucky who promptly named their campsite Lexington. That site is now an important Kentucky city.

On May 25, five weeks after the battle of Lexington and Concord, while the two ships with news of the battle were still racing across the Atlantic, the British frigate *Cerberus* arrived in Boston Harbor carrying three major generals—William Howe, Henry Clinton, and John Burgoyne. In Greek mythology, Cerberus was a three-headed dog that guarded the doorway to hell. Did any of the three generals ever wonder at the irony of that name? On one matter they all agreed: Gage was an incompetent.

Major General Sir William Howe was many things: bon vivant, gourmet, winebibber, cardplayer, womanizer, capable general, and strangely, a Whig who was opposed to Parliament's coercive measures against the Americans. The forty-five-year-old general was a veteran of the French and Indian War, during which he had attacked the Heights of Abraham under Wolfe. He was the brother of Vice-

Admiral Richard, Lord Howe, commander of the British fleet in the colonies. He was to be responsible for planning the Bunker Hill attack.

Major General Henry Clinton was the only son of a British admiral. He had a colorless personality, was a doubting pessimist, a planner and a tactician, and would criticize General Gage openly.

Major General John (Gentleman Johnny) Burgoyne, fifty-three years old, came from a wealthy Lancashire family. He was handsome, charming, witty, a womanizer, a brilliant cardplayer, a playwright, a member of Parliament, and a courageous cavalry commander in Portugal. In missives back to England, he would also condemn Gage's incompetence.

Concerning their own competence as military officers, they were to lose a war they might have won a number of times.

Philadelphia was in the grip of terrible heat rising in waves off the brick sidewalks. The delegates peeled off their woolen jackets and yearned to remove their itchy wigs. They fanned themselves with newspapers and chased the crowds of tormenting flies that flew in from the stables down the street.

New Englanders, unaccustomed to such temperatures, wanted to move the congress to Connecticut. Pennsylvania's Dickinson feared that this would bring the moderates under the influence of the torrid Massachusetts radicals. He quickly scotched the idea. The delegates spent sleepless nights, unable to escape from the heat. Tempers became short.

Perhaps the most sleepless and short-tempered of them all was Samuel Adams. He was waiting for King George III to react to Lexington. He had been working to reach this moment for fifteen years—since the autumn of 1760—and his patience was growing thin.

In the British autumn of 1760, twenty-three-year-old George III buried his grandfather, George II, and as winter came on, clambered up onto the throne of England. With Britain's acquisition of huge pieces of the French and Spanish empires in America, the Caribbean and India following the Seven Years War and with Britain's absolute mastery of the seas around the world, he was the most powerful monarch since the heyday of the Romans.

Had George III cast his eyes over his vast realm looking for future challenges, his glance would have passed over, without hesitation, the small, stout, business failure approaching middle age in a rumpled red suit and a cheap wig, living in a tumbledown house smelling of malt on Purchase Street, along Boston's waterfront. Only one feature might have slowed the sliding glance of the king: a pair of calculating blue eyes.

Yet it was that person with the seedy red suit and vast debts, not another king, who was about to turn the first half of George III's royal reign into a nightmare of turmoil and crisis. This person would cause the king to raise an enormous army and navy and send it thousands of miles from his palace at Saint James to fight battles that would fill military graveyards, would draw the king's European enemies into battle against him, would drain his treasury of vast sums of money for military stores, would attack with weapons the young king had never heard of before, would permanently reduce his monarchical power, and in due course, would pluck from the royal crown Britain's most valuable jewels—thirteen thriving colonies hugging the edge of a vast new world of boundless and unmeasured wealth.

The challenger who would rock the mightiest throne in all of Europe was a man so incompetent in practical matters he had failed in business three times.

How Samuel Adams, with little more than the clothes on his back, developed the tools of revolution and then used them to wrest power from such a mighty monarch has fascinated historians ever since.

Adams's purpose was simply stated. He regarded England as a sick, stratified, corrupt society guided by the anti-Christ Anglican Church and controlled by an unspeakably selfish king and his peerage, who were grinding great misery into the commoners. England was a nation corrupted by its own wealth, greed, and sophistries—lost beyond redemption.

Adams believed that his native colony of Massachusetts was desperately infected with the terminal social disorders of the mother country—but that those ills would be curable under his stern ministrations. He would rescue Massachusetts from the satanic king and restore it to its original Puritan culture.

He envisioned a return to a one-tier society—no privileged classes, no wealth, no poor, no vanities, no crystal, no silverware, no imported furniture, no horse-drawn gilt carriages, no London fashions. Everything would be absolutely functional; everyone would wear basic black clothing and live in the same type of housing. Everyone would get the same education. Read the same book (the Bible, of course). There would be farming to feed the body and religious exercises to feed the soul. Leadership would be, as it had been, by ministers of the cloth. Only.

In short, Samuel Adams would return Massachusetts to 1650, run by worthy descendants of Cotton Mather and Increase Mather—people like himself. And once reestablished, the restored Puritan society would be locked in amber, unchangeable ever after.

Samuel Adams was not seeking justice. He wanted power—the king's power. And he deliberately set out to get it.

But to win freedom from England, Sam Adams, lacking armies and navies, would fight his battles against the king not outside on the battlefield but inside the minds of men. He would conduct a war of ideas.

The problem was that the tools he needed didn't exist. He had to invent them.

So patiently, day and night for more than fifteen years, he forged the tools of revolution—he was the first man in history to do so. It could be said that Samuel Adams invented the art of revolution.

In the beginning he had only the legacy of his family.

His father had bequeathed to him his hatred for England: hatred for Parliament, hatred for the royal officials in the colony, hatred for the colony's self-preening, overlording privileged class. He also left his son a few useful political skills, an exceptionally good education, and finally, a crushing debt imposed by an act of Parliament.

From his mother Sam Adams inherited religion: rigid, narrow, humorless. Christian. Congregationalist.

He saw the world through the lines of John Milton's epic poem of good and evil, *Paradise Lost,* from which he could quote hundreds of lines of sonorous iambic pentameter. Like Milton, he saw Parliament as sulfurous hell and the war against the Crown as a holy war. The king was cast in the role of Lucifer.

Deacon Samuel Adams was a "malsier" by profession, with a malt house and some slaves who steeped barley in water for brewing into beer. By Boston's standards he was prosperous.

The senior Samuel Adams also "distilled the best rum found anywhere in North America from the finest quality of West Indian molasses" smuggled into Boston by Thomas Hancock's ships. Yet he would not drink or allow anyone in his family to drink. He was a preacher who, at his table every night at sundown, held religious services. Yet gossips whispered that the lodging houses he owned were used for brothels.

Deacon Adams was also a noted polemicist, and he often used his pen to defend colonial rights against the Crown and against the royal governors. He was the author of the widely circulated tract "Boston's Declaration of Grievances against the British Governor."

In addition, he was one of the leaders who attacked the British navy's impressments of New England seamen. New Englanders were constantly incensed against the British navy's callous practice of kidnapping seamen from American ships and snatching citizens off the streets of Boston and pressing them into years of brutal servitude on board British warships. To Deacon Adams—and most other Bostonians—this was a violation of the Magna Carta, a practice the king's men were absolutely forbidden by British law to follow.

In January 1748, when an "insufferably arrogant" British ship captain named Knowles perpetrated a "hot press" in broad daylight right on the streets of Boston, abducting carpenters' apprentices and ordinary citizens alike, the city took to the streets. The rioting lasted for weeks, and before it ended, Governor Shirley himself was threatened, his mansion was put under guard against arson, and finally the governor was impelled to take refuge on Castle Island, in the harbor.

In violating its own constitution, the Crown had created a rage in the twenty-five-year-old deacon's son that left him incensed for the rest of his life. It started him down the path of revolution.

A Boston group led by Deacon Adams, trying to overcome the chronic shortage of hard currency caused by the British government's policy of mercantilism, formed a bank based not on gold or on silver but on land. Paper money was issued backed by the real estate owned by the bank.

Debtors loved it. Farmers, mechanics, small businessmen, all cash poor and struggling in a largely barter economy, could pay their debts with abundant paper. They felt this would raise Massachusetts from the severe economic depression that had been crushing the colony for years.

Creditors—all the powerful businessmen in Boston—hated the Land Bank. They wanted to be paid in hard currency of predictable value, gold or silver, not paper, which was inflationary and would lose its value. Scorning the struggling paper money people as rabble, the creditors went to their business friends in England, who went to Parliament, which passed a law outlawing land banks. The act made the officers of the land banks responsible for the paper money the banks had issued.

The senior Adams was confronted with a debt that could ruin him, and he would spend the rest of his days fighting attempts by royal officials in the colonies to take his business. In particular he blamed his ruin on Thomas Hutchinson, a wealthy patrician businessman of Boston, for bringing about the act of Parliament against him.

Heaping insult on insult, the governor's council—the upper house of the legislature—prodded Governor Shirley to remove the deacon from that chamber even though he had been duly elected. The council had traditionally been the preserve of the province's first families. The deacon, now the victim of upper-class snobbism, was ejected.

Young Adams added Thomas Hutchinson and the colony's upper classes to his hate list. He vowed revenge.

At Harvard, Adams discovered the English philosopher John Locke, who taught that the relationship between a king and his subjects was voluntary, based on justice and natural law. This contract was breakable anytime the king violated it. To save the state, Locke asserted, it was lawful to resist the supreme magistrate. To save the state, the illegal became legal. Adams would use this argument for the rest of his political life. Every citizen had the duty to resist the abuses of tyrannical power—whether it be the economically enslaving mercantilism or the ejection of a duly elected official because he did not belong to the right social class.

After college, Adams was apprenticed to a merchant named Cushing, who recognized very quickly that business was not young Adams's forte and returned him to his father. Samuel Senior next gave his son the very substantial sum of £1,000 to start his own business. Adams made a foolish loan that soon cost him a huge piece of his capital. His father philosophically wrote off the losses and brought him into the malt business.

When he died in 1748, Deacon Adams left his son the house on Purchase Street, the malt house, and the unpaid land bank debt. In time, Adams lost the malt house, the home fell into disrepair, and the land bank debt continued unpaid. Adams had no time for business. He was studying the roots of power, trying to find the road to the throne.

A true democrat, he believed all people in society should live on the same level. No privileges. No lording it over others. Despite his education and social standing, he preferred the world of the tavern and coffee house, talking politics with workingmen, rope pullers, caulkers, blacksmiths, sailmakers.

Adams was developing his first weapon for revolution: It was his single most powerful gift—his ability to make men unhappy. He could stir a placid crowd to anger in minutes. He always delivered the same emotional message over and over to one purpose—to take away the Crown's authority over the colonies and use it for his own purposes. And this he did, day and night, month after month, one year following another. He never gave his auditors rest. He never stopped attacking.

Yet Adams kept one foot in the world of the educated, talking with younger alumni from Harvard. He recruited to his cause young men from various social strata—Dr. Joseph Warren, Dr. James Warren, Dr. Benjamin Church, Paul Revere, John Quincy, and his own second cousin, John Adams of Braintree. He swept them into his inner circle.

He was networking ceaselessly. Exchanging ideas. Planting ideas. Attacking ideas. Nudging everyone closer to his concepts of liberty and social justice. And all this time, he never took his eye off the hated Thomas Hutchinson, whose wealth and power grew apace.

While his malt business was expiring from neglect, Adams was elected one of the six collectors of taxes for the town of Boston and managed to make a hash of that also. The job of tax assessor was not demanding, but he was so indifferent to routine and daily duty that he often failed to collect taxes. Many people discovered also that he was a soft touch for a hard luck story.

Meanwhile, his father's debt still dogged him. In 1758, the General Court, the colony's legislature, authorized the land bank commissioners for the fourth time to auction the Adams estate.

Adams went on the attack. He claimed that authorities already had agreed that he did not owe the money on his father's debts and threatened the sheriff and any purchasers with lawsuits. Then, instinctively finding the emotional issue, he switched the ground of the dispute by accusing the commissioners of injustice for hounding him. Then he put them personally on the defensive: He said they were bleeding the public just to pay their own excessively high salaries.

He attacked the whole idea of property and auctions. If they can do this to me, he roared to everyone in Boston, they can do it to

Samuel Adams

you. Adams had buried the true issue of the land bank debt under emotional issues he himself had created. The auction was cancelled. The land bank commissioners were all forced to resign. And Samuel Adams had forged a new political weapon. He had developed two rules of politicial subverting. First, put your enemy in the wrong and keep him there. And second, find or create an emotional issue.

With the end of the French and Indian War, he watched with narrowed eyes as Boston's merchants gathered the fruits of their prosperity. Each ship from England brought more rewards for the wealthy—the crystal and tableware, furniture, books, clothing, paintings, drapes, gilded carriages that arrogantly clattered through the narrow streets, forcing pedestrians aside.

In particular, living on pennies himself, Adams watched with mounting frustration the growth of his enemy Hutchinson—whose appetite for power and wealth seemed boundless. Hutchinson lived in one of the finest mansions in Boston. His importing business was prospering. Through intermarriage with other families such as the equally patrician Olivers, he was building a dynasty for his children. And the indefatigable Hutchinson was serving in multiple public offices as a judge, as a legislator, and as a government executive. Holding multiple offices was to Adams morally wrong.

In the meantime, Adams taught himself a new art—writing propaganda. At first his style was crabbed, stilted. Writing didn't come naturally to him. But he persisted, night after night by candlelight. He studied the writing styles of others; he learned from the Bible, from polemicists such as Locke and Trenchard and Gordon. Using the technique of pro-con, giving an enemy an inch in order to take a mile, in time he became a brilliant propagandist.

At the same time Samuel Adams was gathering and organizing a collection of waterfront mobs who were controlled by his lieutenant, Will Molineaux, a draper, and on occasion even by Paul Revere. Henceforth, Boston was controlled by a "trained mob" glorified by its title: the Sons of Liberty. Sam Adams was its keeper. Adams fashioned another powerful revolutionary tool when he helped spread Sons of Liberty organizations everywhere in the colonies, where they could be orchestrated into mobs for demonstrations, intimidation, and street violence coordinated with events in Boston.

Sam Adams was not an innovator. He was a copier, an assembler, a refiner with a jeweler's eye. What he lacked was an additional philosophical base for his program, an intellectual juggernaut that could attack the smug and shoddy intellectual structure of the British ruling class.

And one day he found it, spread before him like a display in an Arab bazaar. He found it in the mind of James Otis.

Otis was the well-educated son of a powerful businessman and politician. The opposite in many ways of the serious, determined Samuel Adams, Otis was charming, witty, and warm, a delightful raconteur, a wonderful conversationalist, welcomed with open arms wherever he went. And he did go everywhere, moving with ease through the various social strata of Boston.

A serious poet, a student of history and political thought, an author of books in both Latin and Greek, Otis was recognized as one of Massachusetts's most brilliant lawyers, the match of any in England, and also one of the colonies' most bewitching orators. Without notes, extemporaneously, he could speak for hours, compellingly presenting his case, festooning it with appropriate quotations from

British law and from his extraordinarily retentive memory in Latin and Greek. His talk was filled with new ideas, fresh evaluations, unexpected insights.

Everything he did went to excess. John Adams said of him, "He talks so much and takes up so much of our time, and fills it with crass obsceneness, profaneness, nonsense and distraction that we have no [time] left for rational amusements or inquiries."

His was also a mind that was constantly flirting with bouts of madness that in the end captured him entirely. Strangely, James Otis accurately predicted his own death—by a bolt of lightning.

Despite his erratic mentality, Otis wrought a profound change in Adams's mind that helped bring on the American Revolution. One bitterly cold day in February, 1761, James Otis stood up in court and uttered an intellectual cornucopia that fed Adams's propaganda excursions from thenceforth. Of the profoundest significance, that day James Otis, while creating a legend of himself, gave Adams the philosophical authority he needed.

To catch smugglers, to search for contraband, and to increase the volume of customs duties and fines, the Crown's customs officers were using Writs of Assistance. These writs gave the customs officials unbounded power to go anywhere, search anything, break down any door. Intended mainly to intercept smugglers, these Writs of Assistance could be used against anyone, and they had no expiration date. There was one problem with them. There was a growing body of lawyers who believed they were unconstitutional and should be banned.

Governor Bernard, who by law received a rake-off of one third from each auction of impounded property—and was thereby fingering a fortune—had learned that the Massachusetts' chief justice had come to believe that the writs were unconstitutional. Fortuitously for Bernard, the chief justice died before taking any action. Bernard needed to quickly find a new chief justice who would uphold the legality of the Writs of Assistance. He deftly picked the right man for the job—the perennial office seeker and loyal Loyalist Thomas Hutchinson.

But Hutchinson was a wealthy businessman without a jot or tittle of law education or experience. Predictably, his appointment caused strong objections. This was the very kind of London-style political jobbery that was costing the Crown its respect from the colonists.

But the governor wouldn't back down. There was too much money at stake.

When he was appointed chief justice, Hutchinson was also lieutenant governor of Massachusetts, president of the Council of the General Court (the upper house of the legislature), judge of probate in two different counties, and commander of Fort William.

James Otis felt so morally outraged by the writs that he brought suit against them in the name of sixty-three Boston businessmen. He refused all legal fees.

The trial was one of the great dramatic moments in colonial legal history. On a frozen, slippery, snow-filled day, there was standing room only in the old Council Chambers upstairs in Boston's Town House. Every lawyer who could get there crammed himself into the seats and the aisles and along the walls under the portraits of Charles I and James II.

The entire Massachusetts Superior Court, consisting of five judges, including Hutchinson, filed into the courtroom, impressive in their crimson-and-white robes and shoulder-length wigs.

There to observe the critical proceedings, Governor Bernard and his entire council were already seated. Also seated, in three rows, were the sixty-three plaintiff merchants. At a separate table, with special permission to be there, sat a young law student named John Adams. Adams was armed with a sheaf of paper, six sharpened quills and a pot of ink; he was to write the only surviving record of the historic case.

At two o'clock, Otis stood and fixed his February eyes on Chief Justice Hutchinson and Governor Bernard. Then he spoke hypnotically without a break for more than four hours. Observers dared not cough, shuffle their feet, or shift their chairs. In paroxysms of anger and indignation, Otis often twisted his ruffle, pulling it to threads in tempo to the rising and falling of his voice.

Otis's presence was frightening, his anger intimidating, his pacing threatening, battering the court and the council with his anger. Reviewing the original purpose of the writs—enacted under Charles II in 1663 for use against Dutch ships during a trade war—he reached back to the Magna Carta itself.

He denied that even Parliament could abrogate the rights of private property, the home of any Englishman, be he even the lowliest, most humble fisherman. What was at stake was the right of a man to his life, to his liberty, and to his property.

"This writ is against the fundamental principles of English law!" he asserted. The lowliest man, he stated, should be as safe in his home as a prince in his castle—safe from kings, safe from Parliament.

He stunned the whole court with a veiled threat when he condemned "the kind of power, the exercise of which, in former periods of English history cost one king of England his head and another his throne." This was in direct reference to the two portraits hanging on the wall—of Charles I and James II.

In finishing, Otis emphasized the simple argument that was to haunt relations between England and the colonies for the next sixteen years: "An act against the constitution is void. An act against natural equity is void."

The Writs of Assistance were against the natural equity of the Massachusetts charter, which long predated the writs. Therefore, in violation of the charter the day they were enacted, the Writs of Assistance were void.

Otis sat.

During his speech, darkness had covered the city; the court was now lit by candlelight. It had been an awesome performance in front of the leading legal minds of the colony, and when he was through, Otis was famous. Lawyers, hurrying through the frozen streets, seeking warm fires in taverns, restaurants, and homes, talked of that day for the rest of their lives.

A law contrary to the constitution is void.

Samuel Adams could recite Otis's words with great emotion.

Man's right to liberty and property is inherent, inalienable. Man's right to freedom is higher than the state's right to collect revenue.

In those four hours, James Otis had intellectually lit Samuel Adams's way to the throne.

As for Samuel Adams's cousin, John, fifty years later the elderly ex-president would write, "Here this day, in the old Council Chamber, the child Independence was born."

Chief Justice Thomas Hutchinson, predictably taking the case under advisement, later announced that he was conferring with British lawyers. Finally, months later and to no one's surprise, he decided that the writs were legal. He might have gotten high marks for his decision from Crown authorities avid for their dibs, but he got low marks from lawyers—three years later the highest justice in England declared that the Writs of Assistance indeed were and always had been clearly unconstitutional.

But the damage was done. The customs officials, now free to range through the waterfront, relentlessly impounded ships and warehouses, auctioning everything to the walls and eagerly pocketing their one-third share of the take. The fines the court exacted were staggering. John Hancock alone would eventually owe the British Crown the unimaginable sum of £100,000 in smuggling fines.

Samuel Adams watched the irreparable damage the writs were doing. Boston's economy could not sustain a nine-pence tariff on molasses, which was why ship masters began to smuggle in the first place. Boston's sixty distilleries, a major component in the city's economy, were going to be put out of business. The writs were destroying the colonial economy on which the British economy depended and were disaffecting two and a half million of the king's subjects. No one was benefiting except a handful of Crown officers led by the royal governor. Parliament was working against its own best interests again.

And although Otis's performance had inspired Adams, the Hutchinson decision left him smoldering. If royal authorities could use an unconstitutional law to illegally appropriate money from their own citizens, Adams reasoned that he could use equally unconstitutional weapons to fight back. He believed that all citizens had the duty to resist the abuses of tyrannical power—even if they had to throw every rule of law out of the window. He came to regard Hutchinson, Bernard, and the customs officers as nothing but criminals in office. And if need be he would use criminal acts to attack them.

Samuel Adams's apprenticeship was nearly over. He now had a full-blown plan to wrest freedom from the Crown: He was going to destroy the authority of the royal infrastructure in Boston—particularly that of royal officials, the royal courts, the customs agents, the official Anglican Church, the military, and the navy, as well as their patrician supporters. His primary target was still Thomas Hutchinson—but his real target was the king of England.

Adams now devoted all his time to radical organizations in Boston, including the Merchants Club, the Boston Masonic Society, the Monday Night Club for radical politicians, and the vitally important Caucus Club, which Adams's father had once headed.

Samuel Adams was uncovering and coalescing the voices of dissent.

He also worked the tavern circuit, making the Green Dragon Tavern on Union Street "Headquarters of the Revolution," the place where the Sons of Liberty met to be inflamed by Adams.

Adams also cultivated another potent and highly vocal group of dissenting supporters, those "trumpeters of sedition," the "Black Regiment" of Congregational ministers who hated the established Church of England and who not only preached against but also actively campaigned against both the Royal government and Anglican Church. Every Sunday these ministers rained down on their congregations denunciations of Britain, Parliament, the landed gentry of England, and native Loyalists, sending their congregations away so filled with anger that more than once they burned down the homes of Loyalists.

Adams now had a four-pronged attack in position: a deep well of philosophical concepts from Otis, Locke, Trenchard, and others; a network of radical politicians soon to stretch through most of the colonies; a network of newspapers through which he could push his relentless barrage of propaganda and radical ideas; and his closely controlled waterfront mobs—the Sons of Liberty.

All he needed was a new issue, which the king's cabinet soon provided: the Sugar Act of 1774, which imposed a tariff on molasses, the primary ingredient in Boston's world famous rum.

There had been an existing tariff on molasses of six pence per gallon for thirty years but it had rarely been collected. Molasses was in fact smuggled past bribed revenue collectors routinely. Even though the new Sugar Act lowered the tariff from six to three pence, it called for rigorous enforcement by the beefed up Royal Navy. London, intent on finding a tax to impose on the Americans to help defray the cost of colonial and military administration, failed to understand that an *enforced* tax of three pence would be ruinous to the rum industry in Boston.

Sam Adams was the first and for a while the only voice raised against the new Sugar Act. His objection was not economic. It was legal. The new tariff was plainly taxation without representation, he said. It was against the British constitution, against the Massachusetts charter, against natural law. It was void.

But his timing was bad. For although the Boston merchants were in an uproar over the new act, the Massachusetts farmers were unaffected and thus, unsympathetic, while the other colonies were coolly indifferent. Adams, knowing that the farmer's backing was an essential element in his power base, adroitly turned the tax on rum into a tax on them. He warned them that the next tax—and there certainly would be a next tax, he assured them—could be on land, their farmland, then on farm produce. And they had no representation in Parliament to fight for them. Now the farmers became duly alarmed.

Even before Adams could wrest full propaganda value from the

Sugar Act, the British government gave him an even greater weapon when it passed the Stamp Act the following year, 1765.

This was a perfect broad-based cause for Adams, far more useful than the tax on molasses. The new act required that every document, both legal and public, every will, every law suit, every property transfer, every marriage license—indeed, even every copy of every newspaper—had to carry an official stamp to validate it. Documents without the stamp would be unenforceable under the law. Special government offices were to be set up throughout the colonies to sell them.

From Adams's viewpoint this new Stamp Act brought into his camp two of the most highly vocal groups in the colonies, lawyers and newspaper publishers: both vital to Adams's program, both outraged.

But London felt it was on solid ground because similar stamps had long been required in England where they brought in over one hundred thousand pounds a year into the Exchequer.

Now Adams's timing was perfect. Economic distress was prime feeding ground for any rabble rouser, and hard times in Boston, caused in part by the new tariff, had dampened Massachusetts's entire economy and had spread to other colonies. Even before the new act went into effect, Adams's propaganda campaign in the newspapers had the colonists everywhere furious.

Addressing an ever wider audience already in an uproar over the Sugar Act, Adams now took point blank aim at the new law and fired all four of his cannons.

He mounted a propaganda campaign through the *Boston Gazette*. By now he had learned to focus on a single issue—against the Stamp Act, the issue was taxation without representation.

Adams was the first man in history to learn how to brilliantly use newspapers as instruments of propaganda, and in his heyday he was feeding inflammatory propaganda into all the radical newspapers up and down the coast at will. The growth of newspapers, of course, played right into his hands. In 1763 there were twenty-one newspapers in the colonies; by 1775 there were double that number—forty-two, almost all of them anti-British.

Truth was his first victim. To radicalize the populace Adams had adopted a total disregard for it. In his writings he employed slanderous lies, unvarnished propaganda, and rabble-rousing rhetoric. He whipped the people of Massachusetts and many other colonies into an anti-British fury.

One night, John Adams sat in the Green Dragon Tavern and watched Samuel Adams and *Gazette* publisher Benjamin Edes outline a collection of outrageous stories for the next edition of the *Gazette* and then write them there at the tavern table.

The royal governor blamed the *Gazette* for inciting riots: "The misfortune is that seven eighths of the people read none but this infamous paper, and so are never undeceived. . . . the people hear nothing but what their ringleaders chuse they should hear."

First minister of the cabinet and sponsor of both the Sugar Act and Stamp Act, George Grenville, waved the *Gazette* in the House of Commons and said that Americans were being prepared for rebellion by means of subversive newspapers. When he demanded the offending papers be closed, he was told "it was below the dignity of Parliament to pay any regard to angry newspaper writings, who were as frequent and as impudent here as they could be in any country."

Failing to silence the *Gazette* was a major error. In the propaganda war, the British and the American Loyalists lost the battle of the press long before they lost any battles on the battlefield.

Adams also had turned to terrorism. He would nullify the Stamp Act by eliminating the stamp distributor—the wealthy patrician Andrew Oliver, a relative by marriage of Thomas Hutchinson.

Samuel Adams turned out his first mob. It burned Oliver in effigy, then vandalized with axes Oliver's just-built stamp office, then marched down the road to the Oliver house and stoned a number of windows. The next morning, in terror, Oliver resigned.

Adams was not satisfied. He wanted Oliver's utter humiliation. Asserting that his original resignation was not sincere, the Adams mob now ordered Andrew Oliver to make another resignation at the Liberty Tree on the Boston Common.

In a rainstorm flogged by a sharp wind, a huge crowd turned out to watch the terrified man take an oath of resignation administered by Justice Francis Dana under the vengeful eyes of Samuel Adams. Thus, one of the oligarchs of the colony had been publicly humbled and mocked.

An Adams mob attacked the home of William Story, deputy registrar of the admiralty court, then moved on to the home of Benjamin Hallowell, completely destroying the building.

In the weeks that followed many Loyalists were attacked and beaten.

Samuel Adams then produced his masterpiece of violence. He blamed his arch enemy Thomas Hutchinson for urging Parliament to enact the Stamp Act, which was not true. He then dispatched his mob, which spent the entire night completely destroying Hutchinson's home. Hutchinson and his children barely escaped with their lives. The event was so violent that crowds of people came to see the awesome damage. A great groundswell of sympathy for Hutchinson coursed through the colony, and Adams had to hastily condemn the act in public. He hadn't yet learned how to control the shambling, drunken mobs he had created.

But he had achieved his purpose: He had nullified the Stamp Act. Businesses—even the courts—functioned without them, illegally. The Boston rioting encouraged similar demonstrations in other colonies and provoked violent resistance to the Stamp Act throughout the colonies. Mobs in Rhode Island, New Hampshire, Connecticut, New Jersey, Pennsylvania, Virginia, and the Carolinas forced the stamp distributors there to resign. Except in Georgia, not one stamp was ever sold. Samuel Adams had set the power of the Crown at naught, and more of that power was flowing to him.

To get the absolute power he wanted, Adams employed every terrorist tactic he could think of or copy from others—blackmail, intimidation, house burnings, barn burnings, beatings, tarring and feathering, premeditated riots, political stagecraft like the Boston Tea Party—combined with sophisticated political and legal stratagems.

Hutchinson was later to say of Samuel Adams that there was not "a greater incendiary in the King's dominion, or a man of greater malignity of heart [or one] who less scruples any measure however criminal to accomplish his purposes."

Without adequate British protection, justices and court officials were forced by Adams's mobs to resign from office. After driving many duly elected Loyalists from office, his cohorts replaced them with fellow radicals and thus dominated both the legislative assembly and the executive council that governed the colony. Loyalists' servants were spat upon and made to resign from domestic service. More power to Adams.

With these tactics, Adams had deliberately caused a mass exodus of Loyalists. Open to attacks by Will Molineux's roving mobs, and finding no protection from the British forces in Boston, they had fled, many of them to England. The day of the battle at Lexington, the remaining Loyalists were so terrified they left their homes, abandoning everything they owned, and streamed into Boston and the questionable protection of Gage's British army.

Sam Adams had gotten his revenge on the oligarchs, especially Thomas Hutchinson, who had humiliated his father. He had punished Hutchinson and Bernard for the depredations of the writs. He had brought the possibility of revolution closer. He needed but one more move to completely control Massachusetts. Elective office.

He ran for the assembly. Unemployment was high. Businesses were in trouble. Using the deep recession as a campaign tool, Sam Adams was not only elected; the assembly chose him as clerk, one of the most powerful political posts in Massachusetts. The plebian Whigs had ejected the patrician Tories from the legislature.

At last he had his hands on real power. At last he was on his way.

A minor subject of the king of England in a red suit in a far corner of the empire had the mighty regent pacing in his counting-house, whistling up armies and demanding vengeance.

Governor Bernard, who had enriched himself through the Writs of Assistance, called for an investigation into the riots. The assembly, led by Adams, blocked the investigation. When the governor threatened him, Adams then conducted a sham of an investigation using the very men Bernard wanted to investigate. The resulting whitewash was Adams's thumb in Bernard's eye.

Beaten and recognizing that he was "at the mercy of the mob," Bernard told London that "the power and authority of Government is really at an end."

Parliament sullenly repealed the Stamp Act the following February.

To spread the revolutionary contagion outside the narrow streets of Boston, Adams conceived of the brilliant vehicle of Committees of Correspondence, which quickly sprang up in every town in Massachusetts and then rippled throughout all thirteen colonies. These committees could hardly have claimed to be the voice of the majority. Any handful of rebels in any town could form a committee and join the network. But their influence was enormous. They provided Adams with a pipeline through which he could transmit his incendiary messages to all the colonists. Adams then godfathered a system of post riders to carry his seditious ideas to these Committees of Correspondence, who then dispersed them locally.

Everyone knew that Sam Adams now ran Massachusetts.

Each year, he grew stronger. Each year, he dragged, kicked, and bit Massachusetts closer to revolution and independence. With each year, with each turning of the screw, he wrested another chunk of the king's power from him, frustrating him, infuriating him, goring him into one rash act after another, impelling him to work against his own best interests, and luring him into a wasting, expensive war he would ultimately not be able to win.

The stages that led to Lexington were like stone mile markers planted on the shoulder of the roadway for successive generations to read: the Stamp Act Congress, Parliament's repeal of the Stamp Act, the Declaratory Act, the Townshend Acts and their repeal, the colonists' ill-fated nonimportation agreement, the arrival of troops in Boston, the Boston Massacre, the Boston Tea Party, the Coercive Acts, the closing of Boston, the Continental Congress, and finally Lexington.

The acme of Samuel Adams's career was to come on July 4, 1776, in Philadelphia when he signed his palsied signature on the Declaration of Independence, severing the thirteen colonies from England.

By the time General Gage arrived with his army, the king no longer ran Boston. Sam Adams did. He, not George III, controlled the courts; he controlled the legislature; he controlled the contents of the newspapers; and he controlled the contents of people's minds throughout the colonies. Adams could orchestrate his moves against any adversary. Few in England had the faintest sense that things in the colonies were being so fully manipulated by one man. Few of the 2 million people in the colonies knew how much the content of their minds had been shaped by Adams's propaganda—and that included such men as Washington and Jefferson.

Adams might have felt that the mobbism and the political actions accounted for much of his success. His tactics had driven out 14,000 people. But propaganda was probably more important than all the other of Adams's weapons put together.

With it, he won the war of ideas. With it, he won independence.

Yet, in a larger sense, he failed. He never did get Boston back into its Puritan coffin.

As he sat sweltering in that Philadelphia heat wave during the Second Continental Congress, Sam Adams wiped his brow and wondered if King George had heard the news of Lexington yet, and if so what he was doing about it.

London

On May 26, with all sails set, the schooner *Quero,* carrying the Committee of Safety's report on the battle of Lexington amd Concord, sailed into the English port of Southampton.

It had raced across the ocean in the uncommonly fast time of twenty-eight days, all the while diligently scanning the horizon for the *Sukey,* which was carrying General Gage's report. The *Sukey* had had a four-day head start from Boston, and the *Quero* still did not know whether it had reached England first.

Without delay, Captain Derby, set off at full speed for London, bearing a large printed batch of the Committee of Safety's report.

Two days later, on May 28, the captain reached London, by far the largest city he had ever seen—and the most disorderly.

The British capital was a place of unending violence and appalling poverty, starvation, depravity, crime, brutality, filth, disease, drunkenness, and despair.

Prisoners in Newgate starved to death. The insane were chained in filth in Bedlam and mocked by crowds of bystanders. In Charing Cross, not far from Benjamin Franklin's residence on Craven Street, one could stroll along an array of pillories, endlessly stocked by London's superabundant supply of criminals. Time in the pillories, in which the human body was locked in agonizing postures, was often used as punishment for the most minor crimes. On occasion, if the criminal or the crime was exciting enough, sporting mobs came to see and stayed to stone to death. Few would stop them.

Every six weeks, the courts would put on a spectacular show on Tyburn Hill. Crowds would turn out to see the public hangings. Men, women, and children alike came to see the dangling corpses of men, women, and children. There were so many crimes punishable by hanging that whole days were filled with one execution after another. It was commonplace for a boy aged ten to die on the rope for stealing a penknife, a girl of fourteen for a handkerchief. Tyburn Hill was infamous around the world.

Capital punishment was so common, so frequent, that George Selwyn, a fashionable libertine with few other accomplishments to put on his tombstone, could become a "connoisseur of hangings." What would particularly draw the attention of Selwyn was an execution for treason. Within months of the fighting at Lexington, British judges in Ireland pronounced this sentence on Irish rebels:

> You are to be drawn on hurdles to the place of execution, where you are to be hanged by the neck, but not until you are dead; for, while you are still living your bodies are to be taken down, your bowels torn out and burned before your faces, your heads then cut off, and your bodies divided each into four quarters, and your heads and quarters to be then at the King's disposal; and may the Almighty god have mercy on your souls.

To pass through the streets of London, Captain Derby would have to step around, over, or on excrement from slops, open sewers, rotting animals, and human corpses, dead from disease, starvation, murder, or alcohol. Shortly after Dr. Samuel Johnson arrived in the city, the noted author, lexicographer, and conversationalist described London as "a city famous for wealth, commerce and plenty, and for every kind of civility and politeness, but abounds with such heaps of filth as a savage would look on with amazement."

The artist William Hogarth's depiction, in widely distributed prints, of Gin Lane showed scenes shocking but familiar. Actual counterparts were on London streets everywhere. For those who found urban life unbearable, it was said that the fastest way out of the city was a fifth of gin. Death by drinking was all too common.

The author of the classic novel *Tom Jones,* Henry Fielding, who drew on his experiences as a Bow Street magistrate to limn his eighteenth-century scenes, opined that if imbibing alcohol continued at the present rate, there would soon be few poor people left to drink it.

Fortunate were those who escaped the horrors of life on London's streets by being shipped to the colonies—surplus cutpurses, bankrupts, and prostitutes, some of whom, like Magwich in Dickens's *Great Expectations,* found wealth and happiness there.

In the eighteenth century, England dumped on America perhaps as many as 40,000 convicts. Dr. Johnson thought of the colonies as little more than prisoners' plantations with no rights of protest: "Sir, they are a parcel of convicts, and if they get anything short of hanging they ought to be content."

Ben Franklin jokingly offered to exchange American rattlesnakes for British convicts.

For the great mass at the bottom, life was so unbearable that rioting was the only rational reaction. Riots in England during the last half of the eighteenth century were almost commonplace and almost endless. Riotous mobs regularly roamed the streets seeking violence. Their particular targets were the wealthy. They surrounded carriages and overturned them, beat the occupants, robbed them, looted shops, and routinely broke into the houses of the rich. No one offered a solution. Mobbing was just part of city life. Everyone suffered from it; everyone hated it. Yet the very suggestion of a full-time professional police force was regarded with great suspicion.

In the late 1760s, several bad harvests and higher food costs drove pinched workers from hunger to rage to riot. Troops scattered them. But food prices did not come down. When the firebrand Wilkes was imprisoned, mobs flowed through the streets of London to the King's Bench Prison, intent on releasing him. Hastily summoned soldiers, using their muskets and bayonets, left six people dead and fifteen wounded. Striking watermen then attacked the Wilkesites. Wilkes remained firmly locked up.

When the leading merchants of London planned a rally in support of the king, a mob appeared leading a hearse that exhibited drawings of royal troops shooting unarmed civilians. Standing on top was a man dressed as an executioner. When the mob attacked, the merchants fled.

The king himself, on his way to Westminster one day, was hit by a thrown apple.

Benjamin Franklin described life in London this way to his Philadelphia friend John Ross:

> Even this capital, the residence of the King, is now a daily scene of lawless riot and confusion. Mobs patrolling the streets at noonday, some knocking down all that will not roar for Wilkes and Liberty; courts of justice afraid to give judgments against him; coal heavers and porters pulling down the houses of coal merchants that refuse to give them more wages; sawyers destroying sawmills; sailors unrigging all the outward bound ships and suffering none at all to sail until merchants agree to raise their pay; watermen destroying private boats and threatening [Thames] bridges; soldiers firing among the mobs and killing men, women and children, which seems only to have produced a universal sullenness, that looks like a great black cloud coming on, ready to burst in a general tempest . . . while the ministry, divided in their counsels, with little regard for each other, warned by perpetual oppositions, in emotional apprehension of changes, intent on securing popularity in case they should lose favor, have for some years past had little time or inclination to attend to our small affairs [in the colonies], whose remoteness makes them appear still smaller.

In another letter, Franklin reported that he had witnessed "riots in corn, riots about elections, riots of colliers, riots of coal-heavers, riots of sawyers, riots of Wilkesites, riots of government chairmen, riots of smugglers."

Unpopular ministers could expect a visit from a mob at any time, particularly after passing or rejecting a controversial law.

When Lord Bedford impelled the House of Lords to reject a bill raising the tariff on imported Italian silks, many unemployed weavers in Spitalfields were also impelled—into the worst London riots in years. Some 4,000 of them marched on the king's palace at Kew. The king himself was followed by a large mob. Furious faces hissed at Lord Bedford, and murderous hands pelted his carriage with stones, cutting his hands and bruising his temple.

Bedford's house in Bloomsbury Square was mobbed. The indefatigable diarist Horace Walpole noted: ". . . rioters . . . began to pull down the wall of the court . . . a party of horse appeared and sallying out, while the Riot Act was read, rode around Bloomsbury Square, slashing and trampling on the mob, and dispersing them . . . the house was saved and perhaps the lives of all that were in it.'' A guard of 136 soldiers barely managed to protect the Bedford family inside.

Still in a rage, the same mob threatened to pull down every mercer's shop in London. Seeing political shenanigans behind this, Lord Grenville, progenitor of the hated Stamp Act, asserted that the mob was well disciplined and had obviously been paid for its work. A few years earlier, Grenville himself, on his way to the House of Lords, had been attacked by a mob, and he claimed that the attack had been instigated by two political adversaries, the Duke of Newcastle and First Minister William Pitt. But he had never produced enough evidence to substantiate the charge.

Lord North, head of the king's cabinet, was on his way to Parliament when a mob attacked him, destroyed his coach, and was intent on murdering him when he was saved by two opposition Members of Parliament.

In 1785, after coaxing his unpopular shop tax through the legislature, the younger Pitt abandoned his official residence in Downing Street to elude a mob vowing to have his blood.

Failing to understand the underlying despair that drove this violence, George III spoke for his entire class when he placed obedience to the law above all other considerations—and sternly put down the rioters.

Having reached London, Captain Derby went immediately to the offices of Arthur Lee, London agent for Massachusetts, who was serving in the post Franklin had recently vacated. The captain discovered that he was indeed first with the news of the fighting, for when Lee heard it, he was flabbergasted. In turn, Lee led the captain to a hastily summoned cabinet meeting headed by Lord North himself.

When Captain Derby gave the cabinet a verbal summary of events at Lexington, the cabinet's members reacted as though they had been struck by a thunderbolt. With copies of the report in front of them, including the hundred affadavits of eyewitnesses, two of them by British officers, as well as lurid, highly exaggerated clippings from the *Salem Gazette* and the captain's own testimony to verify it all, they seemed unable to take in the enormity of the news. One London newspaper reported: ''Lord North when he received the unhappy news to government, that the Provincials had defeated General

Gage's troops, turned pale, and did not utter a syllable for some minutes."

The cabinet was reluctant to put any blame on their own colonial policies. The reason for the failure, they concluded, was the choice of the general—Thomas Gage. The blame shifting had quickly begun.

The cabinet decided that the king should be informed immediately.

Meanwhile, in accordance with specific instructions from Dr. Warren, Arthur Lee quickly dispatched copies of the report to political leaders in a number of other English cities, as well as to the Lord Mayor of London, aldermen, and members of the Common Council of London. Soon all of England would know the news—as shaped by the American radicals. Dr. Warren had scored an enormous propaganda victory.

King George III was spending the weekend with his family in the White House at Kew, the royal country palace, and to that place went Lord North and Lord Dartmouth—his stepbrother, his fellow cabinet member, and the secretary of American affairs. The astonished king went through the report with them.

George once again saw that a meeting of minds between the colonists and the Crown was impossible. And it had always been impossible—from the first day he had sworn his vows and ascended the British throne.

In 1760, on his coronation day, George III inherited the greatest power since the heyday of the Roman Empire—and the gravest of problems. Bequeathed to him by his largely indifferent grandfather was an unfinished world war against France, an exchequer that was nearly empty, a war debt that was numbing, and a war settlement that would give Britain enormous new territories that it was not yet equipped to govern. Even larger geographic acquisitions were in the offing. Furthermore, the country was on the threshold of the Industrial Revolution, which was already causing the greatest social convulsions in England's history, which were soon to grow worse.

A firm hand was also needed in dealing with Parliament, where the Whigs and Tories were brawling over the postwar direction the country should follow.

In addition, Crown officials in America were warning of a growing fractiousness among the inordinately valuable American colonies. Matters there would require attention soon.

To deal with all this, Britain needed the greatest king in its history. It got George III.

In his royal ermine and holding his gold scepter, twenty-two years old and a bachelor, the new king was an innocent, cloistered, totally unworldly, insecure, mother-dominated stripling with not one minute's experience in governing.

George III was not only very young; he was immature for his age. He had spent his life mewed up in the palace, where he had been badly educated by his domineering widowed mother and her lover, Lord Bute, who had poisoned the boy's mind for his own political purposes.

From his earliest years, George's self-confidence was continuously teeter-tottering. He hadn't learned to read until he was ten. All that he knew of the outside world had been gained from glimpses through palace windows. He had grown up in extreme loneliness, yearning for affection, the product of a badly fractured family—the bickering German Hanovers.

George I had hated his son, George II.

George II hated his son Frederick, the Prince of Wales. Perhaps *hate* was too mild a word. He and his queen, Caroline, were overjoyed when Frederick died before succeeding to the throne.

Queen Caroline said that Frederick was "the greatest ass, and the greatest liar, and the greatest canaille, and the greatest beast, in the whole world, and I most heartily wish he were out of it." On her deathbed, she said, "At least I shall have one comfort in having my eyes eternally closed—I shall never see that monster again."

In truth, Frederick was a dissolute weakling who would have made a very troublesome king. At his death he left but two things to his father: a new successor, Frederick's son, George III; and gaming debts of such abandoned magnitude that George II, shocked, refused to honor them.

George III was to continue the family tradition by hating his son, George IV.

On the day of his coronation, which was witnessed by a twenty-three-year old future adversary, John Hancock, the new king had already displayed the characteristics that were to dominate his reign.

His most serious personality impediment was the Hanoverian

streak of mulish stubbornness, which caused him to cling to policies long after they were seen to be wrong. Despite his extreme inexperience and innocence, he was sharply opinionated on almost every subject. Worse, he did not willingly accept counsel from his more experienced advisers. He also had a fatal flaw for picking the wrong people; he made calamitous choices both in the government and in his royal family life.

On his list of virtues, he was extraordinarily hardworking, diligent, religious, frugal, and moral. Prudish, some said. Rigid, others said. To those around him, including his ministers, he was a generous, kindly man with a gentle sense of teasing humor. All agreed he pursued his goals with a dogged and even frightening persistence.

The young monarch had grown up dreading the prospect of being king. Yet, accepting it as a sacred trust and duty, he vowed to reign with all his might and main even if it killed him. God had appointed him king and he must do his best.

As head of the Anglican Church, and deeply religious, George dutifully said his prayers from the official Anglican Bible—but with a significant change in the text. In his personal copy of the Bible, he ordered every reference to "our most religious and gracious king" to be deleted and overprinted with "a most miserable sinner." His sense of religious propriety was extreme: He was offended when he learned that Lambeth Palace, seat of the Anglican Church, had been the site of a ball.

The king believed in fresh air, exercise, and an almost vegetarian diet. He drank but little and, with his ponderous uncles as bulky examples, had a horror of inheriting the Hanoverian tummy. Between breakfast and dinner, he allowed himself one slice of bread, buttered, and a dish of black tea. He was given to long walks around his country palace at Kew.

His personal morals were at a sharp variance with those of the grand English pashas on their estates. The mature George was not given to chambering, wenching, drunkenness, or coarse behavior. He set a proper example for his family and country at all times. He was a devoted husband and father, rising at six A.M., to spend two hours every day with his wife and fifteen children.

He was fond of music. He played the harpsichord and the flute and essayed the violin. He favored Handel and twice invited young Mozart for court performances. While composing his London symphonies, Haydn found the doors of the court amiably open.

The king liked theater and encouraged it by inviting celebrated actors such as David Garrick and Mrs. Siddons to give readings at Buckingham House.

George III was also an important patron of the arts; he founded the Royal Academy in 1768. He loved the paintings of American artist Benjamin West, made him historical painter to the court at an annual fee of £1,000, and was twice painted by him.

At heart, King George III was an English country gentlemen. His interest in agriculture and the new techniques for increasing crop yields earned him the nickname "Farmer George."

Most of what he learned about the outside world was carried into his throne room on the words of others—advisers, ministers, ambassadors, bishops, military men, courtiers, sycophants. Much of it was either inaccurate or false.

He knew the vast bulk of his subjects—the lower classes—only through the palace staff of domestics and servants, and he little understood the causes of the violent riots that punctuated his reign and that he often sternly put down.

He knew even less about his American subjects. When former Massachusetts Governor Thomas Hutchinson first arrived in England in exile the year before, he spent several hours in conference with George, who revealed in the process "an abysmal lack of even the most rudimentary knowledge of his distant dominion."

George III was like a blind man who knew only what others chose to tell him or what he chose to believe. He was trying to control a far-flung realm he knew only by hearsay.

With all his limitations, his eyes boxed about by the ruling-class attitudes that led him to false conclusions and historical blind alleys, he genuinely tried to solve the problems that were dumped at his feet in the throne room.

The English historian Christopher Hibbert pictured the king in a few incisive sentences: "An honest kindly man with a most obstinate sense of duty, ill advised by his mother and . . . the Earl of Bute. It would take years for him to undo the damage that Bute did to him."

Neither his defenders nor his detractors—including the Americans—had a true picture of George III. He was neither great nor wicked. Yet upon taking up the scepter, he had no world vision, no plan, no grand strategy to implement, no idea how to solve the great problems he faced. As a new king he was alone and badly frightened.

Aside from the poisonous utterances of Lord Bute, young George was totally bewildered about the true scope of his powers.

Perhaps it is fortunate that George III's father, Frederick, the dissolute Prince of Wales, died before becoming king and before he could teach George III his destructive habits. The only guidance George III had were the ever-intrusive words of his domineering mother Augusta: "George, be a king!"

There was no kingly manual or guidebook to help him. The term constitutional monarchy certainly meant something far different from what it had meant to his predecessor—his grandfather George II—who, speaking very halting English and frustrated by the constitutional limits on his power, spent much of his time in Germany. To his British subjects, he was regarded as "both stupid and obstinate."

Going back even further, constitutional monarchy had had even less meaning for his German-speaking great-grandfather, George I—equally stupid and obstinate—whose only means of communicating with his English ministers was in schoolboy Latin. George I lived in Germany and rarely visited England.

Under George I and George II, power had been indifferently wielded, handed off to others. The Whigs had simply taken it and used it, often with great success. But for the determined young king, getting it back from Parliament was to be another matter.

Lord Bute, instead of preparing the young prince for handling the reins of the realm, went about endlessly stirring up trouble by pursuing his own self-serving agenda. In 1758, two years before George assumed the throne, Lord Bute tried to capitalize on his role as royal tutor when he sidled up to the powerful Whig leader Pitt and whispered of his desire for a more important role in government. But Bute was a Scot. Among the xenophobic English gentry, that was almost as bad as being a Frenchman. Pitt brusquely dismissed the idea as absurd.

Brooding and vengeful, Bute paid back the insult by teaching George to hate Pitt as well as the entire Whig Party, accusing it of fifty years of systematic theft of power from the Crown. George's first order of business would be to fire Pitt and eject the Whigs from power—and, of course, to name Bute head of the cabinet. Bute bided his time.

Like other kings before him, George III soon surrounded himself with sycophants who sighed lies and smothered the truth. But slowly, under the tutelage of past masters at parliamentary politics, he fumblingly learned how to take up and use the sullied weapons of power and to work within the limits of his constitutional monarchy.

During his coronation year, the young king's mother required him to marry a princess—any princess, so long as she was from one of the German principalities. As though ordering a thoroughbred horse, George III sent an army colonel to find him a wife.

To that end, Colonel Graham went to the Continent and there chose a fifteen-year-old girl, Princess Charlotte Sophia of the poverty-stricken principality of Mecklenburg-Strelitz. She was "amiable of temperament," possessed of one Sunday dress, and fecund.

George III tried to conceal his disappointment when he finally saw her; she was short, thin, with a large nose and a large mouth. Within two weeks of her arrival in London she was married, crowned queen of England, and soon became pregnant. She went on to bear George III a child a year—fifteen in all.

George III's fatal capacity for choosing the wrong people was particularly disastrous within his own family. With his father, Frederick, dead, George III was also head of a family of eight younger siblings—obstreperous and unruly, full of rivalries, all jealous of George's new wife—a bubbling pot constantly stirred by George's conniving mother, Augusta. For each of his troublesome siblings, the king had to make seemly provision.

The choices of mates his siblings made were inept enough, but the choices that George III made for them were disasters. These closely paralleled George's equally obtuse choices of ministers.

One brother, the duke of Cumberland, without the king's permission married a notorious flirt, a widow with "the most amorous eyes in the world and eyelashes a yard long: coquette beyond measure, artful as Cleopatra and complete mistress of her passions."

Another brother, the king's favorite, the duke of Gloucester, in 1766 also married a widow, heiress to an earl's fortune. She was hardly a suitable consort for a royal duke. The illegitimate daughter of a milliner, she had cadged a fortune by marrying the senescent Earl Waldegrave. Gloucester concealed the marriage from the king for eight years.

Feeling abidingly betrayed, the king never permitted the duchess of Gloucester to set foot in his court for fear that by inviting her he would scandalize all the crowned heads of Europe and bring disgrace on the house of Hanover.

But nothing was more disastrous for the Hanover family than

George III's incredible choice of a mate for his quiet and withdrawn fifteen-year old sister, Caroline Matilda. In 1766 George married her off to Christian VII, the seventeen-year-old king of Denmark and Norway.

On the genealogical charts, Christian VII was a good catch. His mother, Louisa, was sister to Frederick Lewis, the deceased Prince of Wales, father of George III and Caroline Matilda. That made Christian a first cousin to Caroline—and to George III himself.

But the personal life of the youth had been an indescribable horror. His education was lamentable. Worse, he had been terrorized by a brutal governor, then seduced and debauched by corrupt pages. He had been raised as a semi-idiot.

The marriage was a scandal from the wedding day. Sir Joshua Reynolds, the great British portrait painter, remembered that he could not catch Caroline Matilda right because she sat endlessly weeping. Christian VII, as George III should have known—the rest of Europe seems to have known it—was a notorious bisexual pedophile, a philanderer, and an alcoholic debauchee who habituated the worst stewpots in Europe.

He quickly infected his teenage bride with a venereal disease that soon affected her behavior. From a shy bride, she became noisy, obnoxious, and unpredictably erratic. She dressed only in a black riding habit and strode about under a broad black slouch hat. Eager to cure her affliction, she turned to John Struensee, a German doctor. Struensee, who was also Christian's doctor, controlled affairs of state in Denmark from behind the scenes. Struensee apparently could not cure her disease but he did arrest it. Lonely, neglected, and still more child than woman, she found further consolation in Struensee's bed, a fact she heedlessly flaunted. All over Copenhagen, prints depicting the two lovers were on sale.

The unpredictable Christian VII, now with his half brother as regent, emerged from mental torpor to savage vengefulness, and on the night of January 17, 1772, he arrested both his queen and the doctor. After signing Struensee's death warrant, he then had the doctor tortured to death "with the most extreme barbarity."

Before Christian could do in his queen, George III prepared a naval force to sail against Denmark. The Danes quickly released Caroline Matilda. Deemed unfit to return to court in England, she spent her few remaining years in exile, isolated in a castle in Zell, Hanover, the family's ancestral seat, where on May 10, 1775, she died. She was twenty-three.

Christian VII lived on for another twenty-six years, locked away and under the care of his regent half brother.

George III capped the tragic matchmaking of his sister with another masterpiece a few years later—choosing a bride for his son and heir to the throne, the future George IV. Father and son were afflicted with the same quarrelsomeness and dislike for each other that had characterized the previous two Hanoverian kings—George I and George II.

As the Prince of Wales, the future George IV was given to improvidence, profligateness, and public disrespect for his father. In time, his heedless gambling, drinking, and womanizing put him into the almost unbelievable debt of £650,000.

Parliament, taking measure of their future monarch and finding him wanting, offered to clear off all his debts if he would marry. Prince George, clutching at straws, quickly accepted. And his father, George III, stepped into the breach to make the choice of wife. He was doomed to regret his choice.

Duplicating the method he had used for choosing his own queen, George III dispatched to Germany the earl of Malmesbury to hunt up a proper bride. The earl soon reported back that he had found a prospect, Caroline of Brunswick. She was George III's niece—a first cousin to the young Prince of Wales.

Caroline, said the earl, was a suitable future queen of England. This proved to be an astonishing estimate.

Soon the contracts were drawn and the bride-to-be arrived in England. She stunned the entire court. From her mouth tumbled language used in a German army barracks. She rarely—perhaps never—washed. Nor did she change her undergarments with any regularity, and consequently she "positively stank." To bring himself to marry her, Prince George got himself royally drunk. The marriage was a failure and a future scandal.

This was a bad portent. For as in his family life, also on the throne, George III made calamitous choices. And of them, the two worst—the mischief maker Lord Grenville and the disastrous Charles Townshend—were to cost him his prized colonies.

George III was hardly more than twenty-five when he gave the office of Chancellor of the Exchequer to George Grenville. And

Grenville gave George the Molasses Act, followed by the Stamp Act, followed, many said, by the American Revolution.

George Grenville was a common scold and a moralizing troublemaker. He knew what was right and saw wrongdoing everywhere, except in the mirror.

He got it into his obstinate head that the colonists needed to be brought to heel and that he was the one to do the job. No one knew better than Grenville that the Stamp Act he had cooked up was designed primarily not to produce any meaningful revenue but to force the Americans to accept parliamentary taxation. Fit them with bridle and bit. Bring them to obedience.

No doubt that idea was what first attracted George III to him. But two of Grenville's primary characteristics were two of his greatest vices: frugality and talk.

Money was Grenville's god. Hating to part with a penny, he lived on his investment income while saving his salary. He had, said George III ruefully, "the mind of a counting house clerk."

Grenville was quite aware that the Seven Years' War had left Britain with a great war debt. Yet he failed to balance fiscal efficiency with military preparedness. After the Peace of Paris in 1763, his cheeseparing economies reduced the royal army from 120,000 to 30,000 and reduced the entire royal navy to scarcely seventeen battle-ready ships. With not a penny for maintenance, the ships of Britain's great navy were left to rot in their moorings.

London's newspaper, the *Public Advertiser,* sniped at Grenville's string saving. Mr. Grenville believed, it said, "a national saving of two inches of Candle a triumph greater than all Mr. Pitt's Victories."

As for his relentless loquaciousness, talk was the flaw that ultimately brought him down with the king. The acerbic diarist Horace Walpole, while acknowledging that Grenville was the "ablest man of business in the House of Commons," postscripted this with some acid by observing that Grenville's favorite activity was talking and "brevity was not his failing."

Grenville simply wouldn't take yes for an answer. Long after he had won others to his point, he kept on talking. And talking. "When he wearied me for two hours," said the king, "he looks at his watch to see if he may not tire me for an hour more."

A lawyer by training—Eton, Oxford, Inner Temple—and admitted to the bar at twenty-three, Grenville was twenty-nine when he entered Parliament in 1741, representing until his death his family borough. And like many younger sons in the world of primogeniture, he watched the family title go to his eldest brother without having any hope of inheriting a title of his own. He sought power elsewhere in government.

Colorless, humorless, relentless, he mounted the political ladder by coming to work every day—a most unusual practice—and, in a world of upper-class goldbricking, worked furiously for twenty years. At the age of fifty-one, he at last joined the king's cabinet to finally become prime minister.* "So it was that a man of little mind, blinkered judgment, obstinacy, and bad temper came to be prime minister and set his nation on a course that would cost it dearly."

Grenville was not just frugal. He was also greedy. Squeezing his own farthings, he scooped his hand into the public pot to fill his family coffers. Yet when the king wished to give sinecures to his younger siblings, according to the practice of the day, Grenville gave him a lecture and turned him down. George sought elsewhere for his sinecures and never forgave Grenville.

In the end, Grenville was toppled by the abiding flaw of all scolds: Although he could and would admonish others at length, even the king, he could not—would not—allow others to scold him. When convinced he was right—and that was nearly always—he exhibited his worst failing: deliberate deafness, a refusal to hear advice and criticism from others. And it was that obstinate self-righteousness that set the colonies on the path of parting from England.

When Grenville decided to put through his notorious Stamp Act, others came forward to deter him. His own secretary was one. Richard Jackson, a man worth listening to, was a Member of Parliament of some stature, a well-off merchant, and a fellow lawyer. More to the point, Jackson was also the agent for Connecticut, Pennsylvania, Massachusetts, and New York at various times. No one in government could speak with more authority on the temper of the colonists. Jackson warned Grenville that the Americans would react strongly to any tax imposed from London—sugar, stamp, or otherwise.

Adding to the chorus were the voices of other colonial agents, as well as the royal governors in the colonies. They told him bluntly: Don't. For their pains, they all found themselves confronted with Grenville's "cloud of indifference."

Furthermore, even prior to the drafting of the Stamp Act, Gren-

*Or first minister, as it was so designated then.

ville had been advised by the surveyor general of the customs in the northern colonies. Had he listened he would have learned that even if the Americans had willingly complied with the Stamp Act, due to the Crown's mercantilist system the colonies had no hard currency to pay for the stamps. The law, if enforced, would have quickly destroyed the colonial economy and severely damaged Britain's.

Furthermore, Grenville made no provision for enforcing the law or protecting those who supported it. When the Stamp Act led the colonists to retaliate with their riots and their nonimportation agreement, "Americans who stood out loyally in support of the mother country were obliged to face the mobs without protection."

Despite this formidable array of cautions, George Grenville deliberately picked a quarrel with the colonies. Britain had a war debt, and the Americans were going to help defray it or else.

The Stamp Act proved to be unenforceable. The colonies went on a rioting binge, forced the stamp agents to resign, and, with the exception of Georgia, prevented the distribution of a single stamp. The harm done by the Stamp Act to the relationship between the Crown and the colonies proved irreparable.

The end came when William Pitt, now Lord Chatham, and Grenville's brother-in-law, rising from his sickbed, came before Parliament to tell it what it did not want to hear. Roundly damning the Stamp Act, Chatham shocked the whole chamber when he asserted that Parliament did not have the right to tax Americans. He emphatically endorsed the colonies' refusal to accept taxation without representation. Chatham's speech brought about a famous exchange between the two brothers-in-law.

Grenville demanded: "Tell me when America was emancipated."

Chatham responded: "Tell me when it was enslaved."

The king was now confronted with the problem of what to do with a bill that should not have been passed, especially since he did not wish to rescind it.

George III ejected Grenville, and Grenville exited, still talking, still scolding the king. The king, in permanent dread of ever hearing that endless voice again, saw to it that Grenville never got another government appointment. Grenville, resistant to rebukes to the end, died in 1770, thereby evading his share of criticism for the mischief of 1775 that he had helped bring about.

As for the Stamp Act, His Majesty's government had to eat humble pie or fight. Parliament passed around the forks and began eating. A piece of the pie was sent to the king, who was told he had to accept the law's repeal or raise an army. He reluctantly, doubtfully, ate.

Horace Walpole wrote that rescinding the Stamp Act was "setting a precedent of the most fatal complexion." Indeed, Parliament was never to regain authority over America.

George's next mistake was probably his magnum opus: appointing Charles Townshend to the cabinet.

The crisis occurred when Lord Chatham, head of the cabinet and probably the only man who could have controlled Townshend, fell desperately ill, retired to his estate, and stayed away for two crucial years in the grip of "the madness of gout." That left Townshend in control of the worst possible office—Chancellor of the Exchequer—at the worst possible time. Townshend found himself without a leash, free to get into mischief. And he did.

His nickname alone should have been a warning. In the London coffee houses and taverns he was known as Champagne Charlie, "one of the flashiest politicians in England." His main claim to fame: He could make brilliant speeches while totally drunk.

Burke once described Townshend as "the delight and ornament of the Commons and the charm of every private society." Walpole felt that, without his glaring faults, Townshend had the intelligence and ability to have been "the greatest man of this age." And truly his talents were dazzling. Walpole observed that Townshend "had parts that embraced all knowledge with such quickness that he seemed to create knowledge instead of seaching for it." He added that "in him it seemed a loss of time to think." But, with his usual habit of giving, then taking away, Walpole also recorded, ". . . he studied nothing with accuracy or with attention."

Townshend was as inconstant as a summer wind; he was capable, when expedience dictated, of completely reversing himself—even in the same speech. He was too flippant, too arrogant, to bring off his posturing. Many said he was unscrupulous and the Townshend duties indicated that. Others said he was unreliable. His whole history pointed to that.

Scottish historian David Hume, who observed him with grave reservation, said, "He passes for the cleverest fellow in England."

Walpole believed that Townshend's father was mad, and he might have been right. Townshend himself was given to falling fits, which might have been attacks of epilepsy. Determined to operate outside party lines, Townshend vowed he would be guided by only his own judgment. "Judgment," says the American historian Barbara Tuchman, "was his weakest faculty."

Ultimately, one thing canceled out his best efforts and destroyed him: Ambition made Townshend hunger to be prime minister.

He came to his Rubicon when as Chancellor of the Exchequer he suffered defeat on a money bill. It was customary to resign when this happened. But Townshend wasn't ready to go. Rather than lose his post and his power, he selfishly created a national calamity. Urgently seeking a job-saving issue, he announced that he had found a way to get money from the Americans to pay for most of their defense and administration. He would tax American imports on glass, paint, lead, paper, and tea. That would shift some of the British gentry's tax burden onto American shoulders. The British land tax could be lowered accordingly.

His plan dumbfounded his fellow cabinet members. This was the very issue the Americans were already howling over. But when several cabinet ministers tried to cite the flaws in his idea, Townshend threatened to resign. The cabinet squirmed. His resignation could bring down the government; in an election, they could lose their cabinet seats. They were no more ready to resign from power than he was. So they gave in to him.

He then went before the astonished House of Commons with his plan. Many were still brooding over the repeal of the Stamp Act and were eager for another confrontation with those damned rebels. But Townshend's program was loaded with flaws. Even if the Americans accepted the act, the hard currency required would have had to come from other parts of the American-British trade balance—to the detriment of the entire economy. Townshend was robbing Peter to pay not Paul, but Peter himself.

Furthermore, it was quite obvious that the Americans would see this bill as another attempt by Parliament to bring the colonies to heel. As such, the Townshend bill was worse—far worse—than the Stamp Act. It was guaranteed to bring on a fight—a fight Parliament was not prepared for and could not win. Here was another bill that would have to be repealed.

Yet on the strength of Townshend's airy promises, Commons promptly reduced the gentry's land tax from four shillings to three, thereby creating a budget deficit of £500,000, which the Americans were expected to pony up.

Both houses eagerly passed the bill without even bothering with a vote count—and certainly without bothering to debate its glaring and obvious flaws.

George III, had he used his quill, could have quickly discovered that Townshend's numbers did not add up; by Townshend's own estimate, the prospective income was only £40,000. His bill was £460,000 short. And more than likely the cost of collecting would exceed the revenue generated. So Townshend's tariffs were threatening to gouge open a huge deficit in the budget while also bringing on a full-scale America rebellion. All this because he did not want to give up his post.

George III failed to stop him, and the Townshend bill was passed.

A few months later, as a grand finale, Townshend died of a fever. He was forty-two. Like Grenville, he would not live to see what he had worked.

The Stamp Act and the Townshend duties underscored one of the principal flaws in the way England governed itself—the families of the peerage, from which were drawn most of the leadership of the country, including Grenville and Townshend.

In 1775, with a population of 5 million people, Britain was run by an aristocracy of less than 190 peers. Chasing wealth and power with a ruthless dedication, these 190 men and their families and followers were increasing their control annually. They did not like—did not allow—challenges by those they considered their inferiors. And colonials were the most inferior of their inferiors. For the ruling class of England, America was as undignified as a quarrelsome servant.

The ruling gentry was a closely inbred group. They went to the same schools, either Eton or Westminster, then on to either Christ Church or Trinity College at Oxford or to Trinity or Kings at Cambridge. After university, they traveled together with their vast entourages on the years-long Grand Tour of Europe. They married off their sisters to each other, were related through blood on both sides of their families—brothers, cousins, uncles, nephews, stepbrothers, half brothers, siblings without end—and further complicated bloodlines with second and third marriages. Fortunes married fortunes, concentrating the nation's wealth in still fewer hands. It would have been difficult to arrange a marriage between a titled young man and woman who were not related to each other in some way.

As a class, for all their privilege, breeding, and wealth, they set consistently low standards for the rest of the nation: "England's upper class delighted in boorish practical jokes, bawdy conversations, brutal behavior, wild gambling, indelicate taste, and heavy drinking."

To maintain their children in comfortable idleness, the gentry had created a great number of sinecures for them, customarily in the ministry, the military, or, for the seriously ambitious, the law—lucrative positions that usually had no duties whatsoever and that were paid out of the government's budget. There were never enough of these posts.

After garnering a typical sinecure, George Selwyn became registrar of the court of chancery in Barbados without ever setting foot on that island. Similarly, customs officials, stoutly refusing to leave the comforts of England, found substitutes in America, starvelings who, surviving on bribes from smugglers, collected little or no customs revenue for the Crown.

The gentry openly appointed each other to additional functionless offices and swapped sinecures like gum cards for their younger siblings, followers, and other dependents.

They even shared the same mistresses. One lady with the appropriate name of Mrs. Armstead was passed from the humorless arms of Lord George Germain to his nephew, the duke of Dorset, who sent her on to Lord Derby, who vacated his place in favor of the profligate Prince of Wales, who watched her end her bedly peregrinations in the mansion of Parliament's firebrand Charles James Fox, who finally gave her what she had been seeking all along—marriage to a fortune.

For income, the gentry relied on land leases, rent rolls, mines, agriculture, huge absentee holdings over the riot-prone Irish, and adventures in India and other exotic places that few could find on the map, all of which grew steadily in value.

Once ensconced in their worlds, the gentry spent their lives far from the violent, crime-ridden cities, visiting each other on their vast and tranquil estates maintained by armies of grooms, gardeners, landscapers, farriers, gamekeepers, farmhands, carpenters, bricklayers, woodcutters, masters of the hunt, and starched domestic servants.

It was an insular world of cardplaying, riding, dancing, eating, drinking, sleeping, and begetting.

Some time after their twenty-first birthdays the titled young men entered the House of Lords for life. There was no training for governmental administration. No standards of performance. No report cards. Levels of accomplishment were generally low. Blue-blooded incompetents—there were many—often hired gifted commoners to work for them. Titled lords, on becoming bored with government, could return home for the hunt and sumptuous balls for six months at a time. Others would run things. Somehow.

Incompetence was as pervasive as corruption. Sir Francis Dashwood, a fashionable fop, served as Chancellor of the Exchequer, which made him responsible for the government's money; yet, as Lord Rockingham, for a time First Minister, asserted, "his knowledge of accounts was confined to the reckoning of tavern bills" while the simple process of adding five figures was to him "an impenetrable secret."

Dashwood achieved his greatest notoriety as the founder of the Hellfire Club, in the ruins of St. Mary's Abbey in Medmenham. Meeting mainly in the summer, mainly at night, it featured obscene orgies of cavorting lords and ladies. The club had another purpose: to parody the rites of the Roman Catholic Church. One night this activity had almost deadly consequences. In the presence of Dashwood, John Wilkes, Lord Sandwich, and others, Lord Orford conducted a mock ritual to call forth Satan from a box. Without warning the box burst open and out leaped Satan himself. Orford, thinking his ritual had somehow worked, nearly went out of his mind in terror. Even when he saw that it was all a hoax—a baboon dressed as Satan—perpetrated by John Wilkes, the unmollified Orford fled.

Despite all the careful intermarriage and the bricked-up social world, despite the privileged positioning, exclusive education, systematic breeding, and awesome wealth, the whole top-heavy "Squirearchy," while producing selfishness, irresponsibility, and greed in abundance, failed in its one justification for being. "Few true leaders," said Barbara Tuchman, "were born in those magnificent country estates."

Despite Parliament's endless self-praise for leading the world in representative government and constitutional law, the entire membership of the House of Commons was actually elected by only some 160,000 voters, while the franchise was denied to millions of people in the cities. Largely rural Devon had seventy representatives in Commons, and remote Cornwall the same, while London, with a population of 700,000, had six. With its teeming tens of thousands, Manchester had no representative in Parliament at all, nor did Sheffield or Birmingham.

Adding to the pervasive corruption of the power establishment, the peers who sat in the House of Lords also controlled the voting in the House of Commons. Each of the great landowners determined who in his own constituency would be elected to Commons. Known as placemen, these handpicked commoners sat obediently in their places and voted as the peers directed them. Most peers controlled more than just one or two seats. In the 1761 session of Commons, the duke of Newcastle controlled forty-nine seats and Lord Bute seventy-six.

Of the 558 men elected to Commons that year, at least 270 also held sinecures as either government employees or government contractors. And, as part of the system, they were paid off continuously with patronage favors or cash bribes from secret government funds.

The king's head of cabinet, Lord North, entered the House of Commons at age twenty-two from the pocket borough of Banbury and held that seat for the rest of his life. Banbury had a total of thirteen voters.

Parliament was "literally and figuratively a private club." And it belonged exclusively to the titled, landed gentry. "Though regularly denounced as corruption," observes Tuchman, "the system was so ubiquitous and routine that it carried no aura of disgrace."

During the general election of 1774, Lord North went about buying seats as though he were purchasing paintings in a gallery. For one group of three seats he agreed to a price of £2,500 each. For others the price was higher. He wrote to the treasury secretary: "Mr. Legge can only afford £400. If he comes to you from Lostwithiel he will cost the public some 2000 guineas . . . Let Cooper know whether you promised £2500 or £3000 for each of Lord Edcumbe's [five] seats."

Like the House of Lords and the House of Commons, the navy and the army were also the property of the aristocracy. The military was a vast racket for younger sons of the privileged. With no military training or knowledge, they bought commissions to get the uniform and the pay, then didn't report for duty for years on end.

The centerpiece of the government's financial corruption was the annual omnibus bill. This covered a huge part of the government's budget, including the entire civil and government payroll. Much of this went for the Crown's bribery of Parliament, for multitudinous sinecures, and additionally for the purchase of government supplies, which included payoffs and bribes throughout the acquisition system.

Another honey pot was the annual military and naval appropriations bill. This included huge sums set aside for kickbacks, rake-offs, and bribes. There was no accounting system of any sort. Tens of thousands of pounds would disappear without a trace.

The earl of Sandwich, First Lord of the Admiralty—"hearty, good humored and corrupt"—conducted one of the most profitable centers of corruption in the entire empire, dealing out appointments, each with a price tag, and taking staggering rake-offs from the purchase of navy provisions. His busy hands—grasping, defrauding, bribing—swept an endless flow of gold into his purse and produced a navy serving rotten meat and weeviled biscuits with barely a dockful of seaworthy ships. As late as December 1774, with the tumult from America alarming everyone, he failed to stop Lord North's ill-advised reduction of the navy's strength by 4,000 men, a fifth of the total.

His depredations were so shocking that the usually unshockable Parliament, finding no navy in the shipyards on the eve of the American Revolution, reprimanded him with a stern vote of censure from both houses.

Gambling was another grave disorder of the rulers that affected their ability to do their duty. In fact, with the peerage setting the example, gambling raged through society with the fury of an epidemic. Advertisements in newspapers would draw crowds to card tables in the homes of fashionable ladies. The wealthy bet their entire fortunes on the turn of a single card. If people couldn't do anything about the weather, they could—and would—bet heavily on it. Incredible bets were made every day at horse races and at cricket matches and dogfights. Farmers impoverished themselves at cockfights. Even the poorest could bet at pitch and toss.

Lord Stavordale, losing the incredible sum of £12,000 with just one throw of the dice, "thanked God he had not played for high stakes." The duchess of Devonshire lost a million pounds at cards and destroyed her marriage.

Self-ruin was not the worst of this addictive gambling craze. Its effect on men in office could be profound. Sandwich was among the most addicted gamblers in all of England. Legend has it that he invented the sandwich to avoid any mealtime interruption of his gaming. One wonders how much of his lyrically large embezzlements of money that should have gone to naval meat, biscuits, and fighting ships went sailing away over the green baize of the gaming tables to feed his compulsion.

Not everyone was amused by this national pastime. By way of criticism, Silas Neville praised the nongambling Americans, who were "much more virtuous and understand the nature of liberty better than the body of people here and have men capable of leading them."

Leadership among the peers was also severely impaired by drinking—legendary carousing carried to the point of debilitation in office. Charles James Fox, a fiery member of Parliament who never said no to a flowing glass, once entertained eight guests during a visit that lasted from three in the afternoon until ten at night. The group drank "ten bottles of wine and sixteen bowls of punch, each of which would hold four bottles." Each man had consumed the equivalent of nine bottles of alcohol.

Yet gaming, drinking, laziness, and incompetence were not the only maladies visited on the country by the ruling class. Much of the disgraceful poverty of the overcrowded cities had been caused by the greed of the landed gentry themselves. When, through agricultural improvements, growing crops became more profitable than grazing sheep, the land barons helped themselves to the common land around the villages. The enclosures drove hundreds of thousands of farm laborers off the land and into the corruption, decay, despair, disease, and starvation of the cities to seek nonexistent employment. Ultimately, between 1760 and 1850, the landed gentry helped themselves to some 6 million acres of land, which they enclosed with barely a nod at the vast army of urban poor they created.

As Britain put its foot on the threshold of revolution, the people with the power to change everything were the people most determined to change nothing.

Adding to the great social turbulence, the Industrial Revolution was already in full flower. And it was burgeoning from the working class, not the peerage. Hargreaves, who had been a barber, Arkwright and Crompton, Priestley and Wedgwood—these brilliant men caused great landslides of change in England, yet none of them had the vote, nor were they fit to be invited to the tables of the country gentry in their vast country homes. Their skills were not only in inventions and machines but in organizing production and sales, creating the factory system. Revolutionaries all.

Such was the society that George III was trying to lead into war. Aside from the gentry, not many wanted to go. Indeed, many were angrily opposed to a war with the colonies, and the king's now legendary stubbornness underwent its most serious challenges from the blunt warning of these subjects. Many voices tried to tell the king that his American policy was working against his own best interests.

In London, within hours of the arrival of the report from Boston, pens all over the city were busy scratching urgent notes, trying to come to grips with the significance of the events at Lexington and Concord. People were slowly realizing a horrible fact: England was at war with its American colonies.

Lord George Germain, First Viscount Sackville, observed: "It is strange to see the many joyful faces upon this event, thinking, I conclude, that rebellion will be the means of changing the Ministry." Germain would soon be in the thick of the cabinet's wartime activities as architect of Burgoyne's disaster at Saratoga.

Colonel Dalrymple, who had led the first contingent of British troops ashore in Boston in 1768 and could talk about those damned rebels volubly, regretted his encounter with the hopping mad Adjutant General Edward Harvey. Harvey roundly scolded him.

"How often," demanded the general, "have I heard you American Colonels boast that with four battalions you would march through America; and now you think Gage with 3000 men and 40 pieces of cannon, mayn't venture out of Boston!"

The *Gentleman's Magazine,* a monthly and one of the most widely circulated of all the London publications, with amazing speed shocked the entire country with the report in its May 28 issue of the affair at Lexington and Concord.

Redcoats had butchered old people and children? Bayoneted women with child? Burned down homes? Surely the king had now gone too far.

The London agent for Massachusetts, Arthur Lee, swore in a letter that the British had fired first.

Stephen Sayre, an American living in London, sadly wrote, "The horrid Tragedy is commenced."

Horace Walpole, fourth Earl of Oxford, Member of Parliament, and diarist, read the patriot account of the battle of Lexington thoughtfully, then, no more pleased than Sayre, wrote to his friend Sir Horace Mann, "So here is this fatal war commenced!"

John Horne Tooke, an English radical with more enthusiasm than good sense, appealed for contributions, "for the relief of the widows, orphans, and aged parents of our beloved American fellow-subjects who, faithful to the character of Englishmen, preferring death to slavery, were, for that reason only, inhumanly murdered by the King's Troops at or near Lexington and Concord." For his efforts, Tooke was given £100 by sympathizers and one year in jail by the king.

Thomas Hutchinson spoke out from his London residence on Brompton Row. American resistance, he felt, would soon collapse. ". . . they must soon disperse," he wrote to the historian Edward Gibbon, "as it is the season for sowing their Indian corn, the chief subsistence of New England."

Lord Germain observed: "The news from America occasioned a great stir among us yesterday . . . the Bostonians are in the right to make the King's troops the aggressors and claim a victory."

John Wilkes demanded, "What are your armies?" He pointed out that the colonies had a seacoast over 1,000 miles long. Two million Americans could raise an army of 100,000. How was Britain to match that and maintain it in the field, 3,000 miles away? The British exchequer would bleed to death between the two shores.

In London, the cabinet's lame efforts to portray Lexington as a victory brought Edmund Burke to the podium, scoffing: "A most vigorous retreat! Twenty miles in three hours! Scarce to be paralleled in history. The feeble Americans who pelted them all the way could scarce keep up with them."

In the House of Lords, the marquess of Rockingham and the duke of Richmond led a brutal attack on the government. "This is a war we can never win!" they shouted. Lord Camden agreed, asserting scornfully that England could not "furnish armies or treasure" enough to subdue all of America.

Other critics shouted that a war with the colonies would draw France into the fray and start the Seven Years' War all over again—this time perhaps with a worse outcome.

Members of the newly formed Economist Club added to the chorus. They condemned roundly the smothering mercantilist economic policies of the king's ministers toward the American colonies as counterproductive and obsolete. The duke of Richmond said: "I wish from the bottom of my heart, that the colonists may resist and get the better of the forces sent against them."

British businessmen were equally upset. Many heard the news in their clubs and expressed dread. As major creditors to the Americans, they had been more worried about the imminent collapse of the colonial economy than an armed rebellion. The king's stubbornly held mercantilism policy, which was very unpopular, had drawn all the gold and silver out of the colonies, leaving colonial businessmen without the hard currency they needed to pay their British creditors. War could ruin an enormous number of British creditors and do staggering damage to the British economy.

Popular support for America was surprisingly extensive. "When the hangman and two sheriffs appeared in the Royal Exchange to burn publicly copies of a journal called Crisis, which espoused, among other things, the cause of the colonists, the fire was immediately extinguished by the crowd, and the officers were pelted with dead dogs and cats."

"More than three hundred ships had been lined up at the wharves between London Bridge and Lime-House with brooms tied to their masts, thereby advertising them for sale by merchants whose business had been destroyed by the colonists' Nonimportation agreements."

Most politicians running for office during 1774 and 1775 endorsed conciliation with America.

Addressing Parliament on the matter of colonial rights, Lord Chatham said with utter simplicity: "Let it be forever ascertained; taxation is theirs, commercial regulation is ours."

Lord Chesterfield felt that trying to fight for a cause no matter how just with an inadequate army was madness. He observed that the grossest folly in government was "asserting a right you know you cannot exert."

Edmund Burke, looking at the upturned faces in Parliament, warned them: "You are rushing to your ruin." He attacked Parliament's ability to "sneak out of difficulties into which they had proudly strutted." He scorned their overweening pride: "They tell you that your dignity is tied to it. . . . This dignity is a terrible encumbrance to you for it has of late been ever at war with your interest, your equity and every idea of your policy." In short, like the king, Parliament was working against its own best interests.

Even the Quebec Act caused a great amount of hostility in London. It gave Parliament the power to tax Canada, to suspend jury trial as was the practice in France, and, horror of all horrors, to permit the practice of the Catholic faith. The abolition of the right to trial by jury was particularly offensive. The *St. James Chronicle* wrote that such an abrogation of a basic English legal practice was "too scandalous a clause to have been framed by any Englishman."

Privately Sir Jeffrey Amherst, former American commander in chief during the Seven Years' War, said he thought the war unwinable. So did General William Howe. And they were far from alone.

"It is that kind of war," Horace Walpole wrote to Horace Mann, "in which even victory will ruin us." And later, ". . . oh the folly, the madness, the guilt of having plunged us into this abyss!"

Thomas Pownall, Bernard's predecessor as royal governor of Massachusetts, "pleaded with British officials to approach controversies with the colonies in a conciliatory and generous frame of mind."

James Boswell regarded the government's policy as "ill digested and violent . . . mad in undertaking this desperate war."

General Edward Harvey warned in terms the king's groom could understand that the Crown was getting into a war it could not win: "To attempt to conquer it by our land force is as wild an idea as ever controverted common sense."

Even John Wesley, the enormously influential founder and leader of Methodism, tried to bring the government around to reason. "Waiving all considerations of right and wrong," he wrote to Lord Dartmouth on June 14, "I ask is it common sense to use force toward the Americans? Not 20,000 troops, not treble that number, fighting 3,000 miles away from home and supplies could hope to conquer a nation fighting for liberty."

From the reports of his preachers in America, Reverend Wesley had a better picture of the American state of mind than the king did. Not peasants, not cowards, the Americans were mentally ready for war. And were preparing accordingly. This would not be a quick, easy victory, Wesley asserted. It might not even be a victory.

The warnings were not new or sudden.

During the vote on the Boston Port Bill, opposition leaders had warned the king that closing the port would precipitate a rebellion.

Member of Parliament Charles James Fox, for one, said so in unequivocal terms: "If a system of force is to be established, there is no provision for that in this Bill; it does not go far enough [which proved fatally true]; if it is to induce them by fair means, it goes too far. The only method by which the Americans will ever think they are attached to this country will be by laying aside the right of taxing. I consider this Bill as a Bill of pains and penalties, for it begins with a crime and ends with a punishment; but I wish the gentlemen would consider whether it is more proper to govern by military force or by management."

During the third reading of the Port Bill, among the objectors was George Johnstone, former royal governor of west Florida, who predicted the most dismal consequences when he said: ". . . the effect of the present Bill must be productive of a General Confederation, to resist the power of this country."

John Dunning, former solicitor general in the administration of the Third Duke of Grafton, saw the Port Bill as heralding "war, severe revenge and hatred against our own subjects."

Burke had bluntly attacked the Port Bill: "Your scheme yields no revenue; it yields nothing but discontent, disorder, disobedience; and such is the state of America, that after wading up to your eyes in blood, you could only end just where you begun; that is, to tax where no revenue is to be found." Burke accused the government of destroying trade worth £5 million a year, an essential component in the economy of England. To get at the radicals such as Sam Adams, he roared, the government was pulling catastrophe down upon itself.

General John Burgoyne, who also held a seat in Parliament, said he would like to "see America convinced by persuasion rather than the sword."

Perhaps the most surprising comment of all came from General William Howe. While canvassing for the election of 1774, Howe had told his constituents that the "whole British Army together would not be enough to conquer America."

In January, just months before Lexington, there still had been time to head off the confrontation.

That month Lord Chatham himself struggled from his estate into wintry London to stand up in Parliament and solemnly warn the king he was bringing on a catastrophe. Barely able to hold himself erect from his crippling gout and uttering every word in great pain, he looked at the crowded chamber in front of him. Not a sound was heard. All eyes were fixed on him as he rose. His compelling voice began by taking them to task for their shortsightedness.

He condemned the passage of the Boston Port Bill because it punished "the innocent trader" along with the "guilty profligates"; it put "a crime of a few lawless depredators and their abettors upon the whole body of the inhabitants." And he warned them yet again: ". . . this country has no right under heaven to tax America."

He noted that the furious American resistance to "your arbitrary system of taxation might have been foreseen." He pointed out that it had produced rioting, nonimportation agreements, the banding together of the colonies in a Continental Congress, but not one penny in new taxes.

Then, fearful that a clash with the colonists was not far off, he solemnly warned that the Crown had to immediately withdraw all British forces from Boston to "make the first advances for Concord." The troops, he charged, were not effective; they were provocative. They might control any area they marched into, "but how shall you be able to secure the obedience of the country you leave behind you . . . ?"

He asked them what size army they thought they would need to subdue the colonists if war came: "What, my Lords, a few regiments in America and 17,000 or 18,000 men at home! The idea is ridiculous. To subdue a region extending over 1800 miles, populous in numbers, valorous and infused with the spirit of liberty would be impossible." To "establish despotism over such a mighty nation must be vain, must be fatal. We shall be forced ultimately to retreat: let us retreat when we can, not when we must."

Sadly, Chatham, ailing and aging though still brilliant as a speaker, had been too ill to do the necessary politicking behind the scenes and didn't have the votes to pass his motion to withdraw the troops.

"The opposition stared and shrugged," wrote Walpole; "the courtiers stared and laughed." Chatham's motion was voted down overwhelmingly, 68 to 18.

Then on February 1, less than ninety days before Lexington, Chatham struggled back into Parliament with a new motion. Again they came to hear him. Again they subsided to absolute silence as he rose. And again he spoke compellingly.

Once more he called for repeal of the Port Bill and the other Coercive Acts, but this time he had an offer to make—a way out of the impasse with the colonies. Instead of Parliament's taxation without representation, Chatham called for a recognition of the Continental Congress as a tax-gathering body for Crown revenues. It was not to be a complete sundering. The Crown would still regulate external trade and would still keep the right to deploy an army.

There it was again. A Plan of Union, America with its own Parliament, just what Galloway was calling for at that moment in Philadelphia, just as Franklin had called for more than twenty years before, just as Parliament would eventually accept for the future Commonwealth.

Lord Gower "rose in great heat" to denounce Chatham's plan. The bill, he cried, betrayed the fundamental rights of Parliament. "Every tie of interest, every motive of dignity, and every principle of good government," he said, demanded "legislative supremacy entire and undiminished." He wanted war.

Defiant, sneering, furious, the majority, wanting to slap down the obstreperous colonists, slapped down Chatham's bill.

Struggling again to his feet to reply, Chatham, his warnings as unbelieved as Cassandra's, now handed them a wall-shaking coda: "The whole of your political conduct has been one continued series of weakness, temerity, despotism, ignorance, futility, negligence, and notorious servility, incapacity and corruption."

With a baleful glare at them, he departed.

Parliament had not heeded a word Chatham said.

Now it was May in London, and all the dire predictions of the opposition had come true. The words of Chatham and Burke and all the others haunted the king's throne room.

At last, on June 10, two weeks after Dr. Warren's report had shocked all of Britain, General Gage's dispatches arrived on the lubberly *Sukey.* In dry language whose understatement must have made it seem as though the document were describing a different military action entirely, General Gage's report did not mention the extent of the casualties. These figures came in a later report.

To the highly agitated cabinet, already filled with stories of British atrocities, Gage's opening sentence must have seemed the words of a madman: "I have now nothing to trouble your lordship with, but of an affair that happened here on the 19th instant."

The focus of Gage's report was on the brilliant rescue operations of Lord Percy, whom he praised highly. He stopped short of declaring that Percy's skill alone had prevented a complete military disaster.

But along with the Gage report came numerous letters from British officers in Boston that soon began circulating in homes and coffee houses and taverns and in the press. These British writers told a completely different story. They stressed the atrocities committed by the savage Americans, magnifying the single braining with an ax into scores of atrocities: ". . . full bad as Indians for scalping and cutting the dead Men's Ears and Noses off, and those they get alive, that are wounded and can't get off the Ground." At the very least, the wounded had their eyes "googed" out. British soldiers, they contended, had set fire only to houses that had contained snipers. Furthermore, the Americans had fired first. And still worse, they did not fight fair, hiding behind trees and barns, loading and reloading behind walls while lying on their backs. The Americans had not fought as upright soldiers but as skulking monsters with scalping knives. Many of America's supporters began to waffle when they read of the American atrocities.

Lord Suffolk, noting that Gage's information was far short of what private letters were relating about the Lexington affair, said: "They don't do much credit to the discipline of our troops, but do not impeach their readiness and intrepidity."

Lord George Germain wrote back to Lord Suffolk with the official viewpoint, complete with official scapegoat: "I must lament that General Gage, with all his good qualities, finds himself in a situation of too great importance for his talents." Germain undercut the general even more by suggesting that his troops might not wish to serve under him any further. Obviously, a better general would bring better results. And the Crown had just sent three of them to join Gage in Boston—Howe, Burgoyne, and Clinton.

The posture was now official: It wasn't the coaches in London who had let the side down. It was the players in Boston.

In considering his response to Lexington, the king seemed to be assessing a different set of facts from his critics.

First of all, he had a completely different view of the Boston Port Bill that had brought on the crisis. As his spokesman during the debate, Lord North told Parliament: "The Americans have tarred and feathered your subjects, plundered your merchants, burnt your ships, denied all obedience to your laws and authority; yet so clement and so long forbearing has our conduct been that it is incumbent on us now to take a different course. Whatever may be the consequence, we must risk something; if we do not, all is over. . . . We must control them or submit to them."

Furthermore, the king did not see Lexington as a defeat. In a note to a flagging Lord Dartmouth, even though he did not have the details straight he was positively upbeat: "I am far from thinking the General [Gage] has reason to be displeased. The object of sending the detachment was to spike cannon and destroy military stores; this has been effected, but with the loss of equal number of men on both sides. The die is cast. I therefore hope you will not see this in a stronger light than it deserves."

But putting a brave face on it could not conceal the dismaying facts: After searching all the corners and closets of his empire he was able to scrape together no more than 40,000 troops, including the Irish infantry. That was hardly enough. He found even less of a navy. And he was, with the greatest concern, watching the behavior of the brooding, wound-licking French king, Louis XVI, who had neither forgotten nor forgiven the drubbing the British had given France during the Seven Years' War. Even George III knew that he could not fight both the Americans and the French.

Aware that it was his own policy during the previous fifteen years that had gotten his country into this impasse, the king did not for a moment entertain the thought of changing that policy. The policy was right; the Americans were wrong.

Indeed, the wrongful cry of the Americans—"No taxation without representation"—could not under any circumstances be accepted. Almost everyone in England was taxed without representation. To give representation to the Americans would mean giving representation to the English. To give representation to everyone was to give away power. And that was impossible. Even the slightest reform could cause great shock waves.

Actually, the king did not feel he had any choice. He believed that if he lost America, he could lose everything else. He could cause the breakup of the British Empire. He continued to be driven by the same fears throughout the Revolution. As late as June 1779, he wrote to the war-weary Lord North: "Should America succeed in [independence], the West Indies must follow them; Ireland would soon follow the same plan and be a separate State, then this island would be reduced to itself, and soon would be a poor Island indeed, for reduced in Her Trade, Merchants would retire their Wealth to

Climates more to their Advantage, and Shoals of Manufacturers would leave this Country for the New Empire."

In fact, independence could bring about a mass exodus of Irish to the American colonies, draining Ireland of its cheap labor. But far worse, in the background there always lurked the ultimate catastrophe. If the very Catholic Irish wrested free, they would eagerly admit the very Catholic French king, Louis XVI, and his troops right into Britain's back door. And peering over the shoulder of the French king were the sullen eyes of the Spanish monarch.

Therefore, for all of these reasons and despite the warnings of his critics, George III wrote a memo to Lord North stating his fixed position: "I am not apt to be over-sanguine, but I cannot help being of the opinion that with firmness and perseverance America will be brought to submission. Old England . . . yet will be able to make her rebellious children rue the hour that they cast off obedience. America must be a colony of England or treated as enemy."

And in a second memo to Lord North he said: "I am certain any conduct but compelling obedience would be ruinous and culpable, therefore no consideration could bring me to swerve from the present path which I think Myself in Duty bound to follow." And: ". . . once these rebels have felt a smart blow, they will submit."

He had made up his mind. Stubbornly, he sent his armies off to fight for eight years in a war that prudent military advice had warned him was hopeless.

Afterward

BOSTON

DR. BENJAMIN CHURCH, of Samuel Adams's inner sanctum, was exposed as a spy for General Gage in the summer of 1775, when one of his coded letters, carried by his pregnant mistress, was intercepted. When he was brought before a court-martial, he pleaded guilty and petitioned for release because of declining health. He was granted permission to visit the West Indies in 1776, on the condition that he never return to his native land. He and his ship disappeared at sea.

DR. JOSEPH WARREN, who dispatched Paul Revere and William Dawes on their rides to Lexington, was killed by a shot to the head at Bunker Hill during the last charge.

MAJOR PITCAIRN was also killed in that last charge, which he was leading. He was killed by a shot to the head fired by an African American militiaman, Salem Prince, who had also been at Lexington and Concord. The major died in the arms of his son, a fellow officer, and was interred under the floor of the basement in the Old North Church. His remains were then ordered shipped to England for reinterment. Only later was it discovered that the wrong bones might have been shipped and that Pitcairn's remains might still be buried in the old North Church's basement or perhaps in an unmarked grave. Major Pitcairn's prized Spanish pistols were carried by General Israel Putnam throughout the Revolution. They are on exhibit in the Hancock Clarke House Museum, a scant few hundred yards from where the major might have fired the first shot in Lexington.

The Provincial Congress gave Major Pitcairn's horse to Concord's patriot Pastor William Emerson.

SAMUEL ADAMS's greatest moment came on July 4, 1776 when he signed the Declaration of Independence. But with the onset of the Revolution, his work was largely done, and his influence in the revolutionary movement steadily diminished. An expert in overturning governments, he knew nothing about building one.

Instead, his career in the Continental Congress descended into factionalism and quarrels with his enemies, particularly with John Hancock, whom he came to hate. He also hated George Washington, and, quixotically rejoicing in his defeats, he was overjoyed by Washington's rout at Brandywine. As a leading member of the Conway cabal, Adams conspired unsuccessfully to replace Washington with General Conway. He fought bitterly against the new constitution, which he saw as an intrusion on the rights of the states. Through an investment in Jamaican real estate, he became prosperous at last. In his diary, William Bentley wrote Samuel Adams's epitaph: "He was feared by his enemies but too secret to be loved by his friends." In his last years, he became, as John Adams noted, a "weeping, helpless object of compassion." He died in 1803 at age eighty-one.

JAMES OTIS, who achieved lasting fame for his brilliant speech against the Writs of Assistance that inspired both Samuel Adams and John Adams, and who had predicted his own death by lightning, went almost completely mad and grew enormously fat from a diet of bread and honey. In March 1783, he wrote a will in which he left to a daughter who had married an officer in the British army a legacy of five shillings. Two months later, he was watching a thunderstorm while leaning on the jamb of an open doorway. A bolt of lightning struck the chimney, traveled across a roof beam and down the wall to the doorjamb, and killed him.

JOHN HANCOCK, like Samuel Adams, achieved the acme of his career when he signed the Declaration of Independence. And as was true for Samuel Adams, thereafter Hancock's stature and influence in the Continental Congress waned. He became an abiding enemy of Samuel Adams, who, he claimed, conspired against him. He was elected governor of Massachusetts, and died in office in 1793 at the age of fifty-six.

GENERAL THOMAS GAGE returned to England, where he continued to be royal governor of Massachusetts in absentia even after American independence, when there was no longer any royal colony for him to govern. Regarded as "a villain in America for doing his job too well and a villain in England for not doing it well enough," for a time he was in military limbo, but by 1782 he was made a full general and died with the requisite honors in 1787. His marriage to Margaret Kemble Gage, who, he believed, had betrayed his plans of attack on Concord to the colonists, was never repaired. She survived him by thirty-seven years, dying at the age of ninety. She never revealed what her role—if any—was.

PAUL REVERE became wealthy. His role in the disastrous attack on a British fort in Penobscot Bay ended in controversy and also ended his military aspirations. With independence, however, he resumed his ruined business and built a thriving copper foundry that had been forbidden under British mercantilism—the Revere Copper and Brass Company. After mastering the science of metallurgy he introduced a number of important inventions. Revere copper sheets bottomed the USS *Constitution.* He had sixteen children and more than fifty grandchildren. There are thousands of Revere descendants. He died in 1818 at age eighty-three. His role in the battle of Concord and Lexington is so celebrated that his reputation as one of America's most gifted silversmiths is often overlooked.

Rachel Revere's note to her husband turned up 200 years later in General Gage's papers. Revere never received it or the money accompanying it. There's no record of what Dr. Church did with the £125, but considering his shaky grip on ethics, one need not be charitable.

Secretary of Massachusetts, THOMAS FLUCKER, Henry Knox's father-in-law, left with Howe's troops for Halifax when Boston was abandoned.

The Fluckers eventually reached England, where he was appointed secretary of Massachusetts in absentia, for which he continued to draw his salary of £300 pounds a year. He died suddenly in England in 1783, the year of the treaty that created the United States of America. He never saw his daughter Lucy again.

Henry Knox, Flucker's son-in-law, was destined to become Washington's brilliant artillery general, earned lasting fame by dragging the cannon from Fort Ticonderoga across Massachusetts in the snow to Dorchester Heights, where they were used by Washington to drive the British out of Boston. Knox later became the first secretary of war of the new United States. An indefatigable correspondent, he left some 20,000 letters after he died in Maine when he choked on a chicken bone.

Former Massachusetts governor THOMAS HUTCHINSON died in England, longing to go home to Boston and be buried beside his wife. During the Revolution his handsome mansion was used by the militia as a barracks. Those furnishings that weren't stolen were sold at auction along with nineteen other Hutchinson properties. The proceeds were claimed by the provincial treasury. General Washington used the Hutchinson carriage during the Boston siege. When the Provincial Congress of Massachusetts published a list of 300 Loyalists to be banished forever, his name headed the list. James and Mercy Otis Warren, who hated him, bought his house in 1781, right after he died.

BROWN BEAUTY, the celebrated mare that carried Paul Revere on his historic ride, turning out the militia in a string of towns and villages in Middlesex County, became exhausted when commandeered by the British road patrol, collapsed, and died on the Concord road.

LIEUTENANT COLONEL FRANCIS SMITH, sharply criticized by several junior British officers for his conduct at Lexington and Concord, received high praise for that same conduct in dispatches from General Gage. At the siege of Boston, he ignored reports of the American artillery on Dorchester Heights that quickly drove the British out of Boston. In the New York campaign, he ignored an opportunity to cut off Washington's retreat. Lieutenant MacKenzie said Smith preferred to look for quarters for himself in that city. Promoted to brigadier, Smith ultimately became military adviser to the king. His portly portrait hangs in the portrait gallery of the British National Army Museum in Chelsea, London.

HUGH, LORD PERCY fought at New York with the same skill and courage he had shown on the retreat from Lexington. He was instrumental in the capture of Fort Washington. Promoted to lieutenant general, he resigned in disgust over the conduct of the war and sailed home in 1777.

ADMIRAL GRAVES, naval commander during the siege of Boston, in a high rage after Lexington, ordered Falmouth, Maine, burned. This caused a great outcry in both America and England and cost him his command in January 1776. He died in 1787. One source describes him as "stupid beyond belief."

LIEUTENANT EDWARD GOULD, whose foot wound at the North Bridge in Concord sent him off careening in a borrowed buggy, was captured and sent to Medford for medical treatment. At Medford, he gave his deposition clearing the Americans of firing the first shot at Concord. He died of his wound not long after.

DR. SAMUEL PRESCOTT, who spread the alarm to Concord when Revere and Dawes were stopped, became a surgeon in the Continental army, then joined the crew of an American privateer sailing out of New England. Captured, he was thrown into prison in Halifax, where he died in agony in 1777. In Lexington, Lydia Mulliken, who became his fiancée on the night of the battle, didn't learn of his fate until war's end in 1783. Eventually she married another and raised a family.

WILLIAM DAWES, the second alarm rider to Lexington, joined the Continental army commissary to fight at Bunker Hill, went into business, did well enough, but died of poor health at age fifty-three. He is buried in the churchyard of the King's Chapel in Boston.

Newspaperman JOHN GILL set up the *Boston Constitutional Journal* in May 1776, after the British evacuation. The following November, Benjamin Edes carried the *Gazette* back into Boston and continued publishing it until 1798, although it declined greatly in the postwar years. Both Edes and Gill died in poverty.

ISAIAH THOMAS kept the *Massachusetts Spy* in Watertown, where, after the war, he prospered with a string of successful papers in other towns, branched out into publishing and bookselling, amassed a fortune, and became a patron of the arts and sciences. An important figure in the history of American journalism, he founded the American Antiquarian Society and wrote the first historical account of American newspapers.

BENEDICT ARNOLD, although revered by many exiled Loyalists, was never fully accepted by the British. He spent his last years, still married to Peggy Shippen, operating privateersmen against the French navy. He died in London in 1801 at age sixty—in debt and, contentious to the end, beset with lawsuits.

GENERAL WILLIAM HEATH was not able to repeat his ring of fire success against Lord Percy in subsequent battles of the Revolution. Elevated to major general, he was active in the New York City area, where in 1777 he attempted to overrun Fort Independence, near Spuyten Duyvil, which was occupied by some 2,000 Hessians. When the Hessians sallied from the fort, Heath's troops fell back in a panic, and a few days later he had to withdraw with the taunting cries of the Hessians following after him. He was reprimanded by Washington. In October 1780, Washington appointed him commandant of West Point, replacing Benedict Arnold, who had defected to the British. With the peace, Heath returned to farming and state politics in Massachusetts. He died in 1814, the last surviving major general of the Revolution.

THOMAS BRATTLE, the wealthy Tory who was chased by a mob of 4,000 that believed he had connived at Gage's seizure of 250 barrels of gunpowder, was never allowed to return home. He wandered for the rest of his days.

DANIEL LEONARD, a lawyer and the author of the *Massachusettensis* papers, was among the 1,000 Tories who left with the British army for Halifax in March 1776. His property and wealth were impounded by the radicals. Leonard took his family to London, was appointed chief justice of Bermuda, and lived to the age of eighty-nine, dying in London in 1829.

JUDGE SAMUEL CURWEN of the admiralty court in Boston, who fled from the radicals and his wife both, did not find happiness in London. He confided to his diary that tarring and feathering would have been preferable to his unhappiness in England. He also found himself increasingly more sympathetic to America as the Revolution progressed. When he returned to Boston in 1783 his wife had an "hysterick fit."

Militia sergeant WILLIAM MONROE, Lexington tavern keeper, lived through the Revolution and had the high honor of hosting the country's new president, George Washington, at dinner in his tavern on the occasion of the president's visit to Lexington in 1789 to commemorate Lexington's involvement in the Revolution.

Dover militiaman JABEZ BAKER, wishing for an unusual garment to make into a scarecrow, had abandoned his plow and marched off to battle. Marching back home, after chasing the British back into Boston, he found the perfect garment for his much-needed scarecrow—a captured military redcoat, made of the toughest British wool and famous for its durability. Baker put it to the ultimate test: he hung the redcoat on a scarecrow frame on his farm. Ninety brutal winters, ninety searing summers later, through frost and rain and burning sun, in 1866, a year after the American Civil War, it was still there.

NEW YORK

JAMES RIVINGTON, publisher of the *New York Gazetteer,* apparently learned a great deal from loose-lipped British officers in his coffee house; he undertook to spy for Washington through the Culper Jr. spy ring in New York City, which gathered information on the British. In 1781 he laid his hands on a set of British naval signals used by Admiral Thomas Graves that were passed on to the French admiral de Grasse. Although George Washington sent officers to guard Rivington, Sears and his gang finally shut Rivington's paper down. His spying wasn't revealed until the twentieth century. His probable reason for doing it: He needed the money. He died penniless in New York in 1802.

Anglican Minister SAMUEL SEABURY, author of the "Letters from a Westchester Farmer" which so enraged the radicals, was brutally arrested by Isaac Sears and his cohorts in front of the terrified Seabury children and was jailed in Connecticut. He later served as chaplain to Edmund Fanning's King's American Regiment. After the Revolution, the clergy of Connecticut elected him the first American bishop of the Episcopal Church. When the English church refused to consecrate his bishopric without the oath of allegiance to the throne, Scottish bishops did.

New York loyalist politician OLIVER DELANCEY, as a brigadier general, became the ranking officer in the Loyalist corps. Bloomingdale, his widely admired mansion, was burned to the ground by radicals in 1778. With all his great wealth—he claimed $390,000—confiscated, he was compensated by the king with $125,000. He died in Beverly, England, in 1785.

The New York LIVINGSTONS could have been the only family with three signers of the Declaration of Independence. Philip did sign it. His brother, William Livingston, who was also a delegate to both the First and Second Continental Congresses, could have signed it—but he felt the document was not necessary and left. He became the first governor of the state of New Jersey (1776–1790). A cousin, Robert R. Livingston, actually worked on drafting it with Jefferson and Ben Franklin but was unable to sign it because he was a delegate to the New York Provincial Congress, which in July 1776 drafted the first New York State constitution. Named first chancellor of the state of New York, Robert later administered the oath of office to George Washington in 1789.

ALEXANDER HAMILTON got into a duel with Aaron Burr in Weehawken, New Jersey, in 1804. Hamilton was killed by the same pistol that had killed his eldest son, Philip, 19, in a duel a few months earlier.

ALEXANDER MCDOUGALL, who fought Captain James DeLancey so doggedly, took that same persistence into the Continental army, where he became a general under Washington.

PHILADELPHIA

THOMAS PAINE returned to England in the 1790s and wrote *The Rights of Man,* a defense of the French Revolution against an attack by Edmund Burke. Accused of treason, he fled to Paris to escape prosecution, became a member of the National Convention there, was

Robert R. Livingston, Jr.

thrown into jail during the Reign of Terror, and, escaping with his life, returned to the United States in 1802. He died in poverty in 1809 and is buried in New York City.

DR. BENJAMIN RUSH, despite his enormous accomplishments in Philadelphia, did not find life all smooth sailing. He was roundly criticized by other doctors for his continued use of bleeding as a medical treatment. He was also embroiled in the Conway cabal to remove George Washington as general. And his attack on his superior military officer, Dr. William Shippen, for poor medical practices in the Continental army left lasting ill will after Dr. Shippen was exonerated of all charges by Congress.

Yet his courage and affection for his fellow humans was unshakable. When most doctors quickly left Philadelphia during the terrible yellow fever epidemics of 1793 and 1797, he was one of the few who stayed to treat the sick. That dedication to public service was finally to prove fatal; he was himself swept away by another epidemic—

typhus—in 1813. He was sixty-seven. No tombstone was large enough to contain all his accomplishments.

JOSEPH GALLOWAY fled Pennsylvania in 1776 and joined the Loyalist forces in New York City. He later returned, guiding the British army from the Brandywine battlefield into the city. When the British left Philadelphia to return to New York he went with them, leaving the city for the last time. He later sailed to England. His wife, who refused to join him in exile, died in Philadelphia while fighting to recover confiscated property for her daughter, who finally reclaimed it through the courts in the 1790s.

The significance of the Galloway Plan has been lost in history. Some speculate that had the Galloway Plan prevailed, it might have changed the course of American history and circumvented the Revolution. There might have been no American Revolution and no French Revolution, the United States might still be part of the British Commonwealth, and consequently both World War I and World War II might have been averted because Germany would not have so brazenly gone on the offensive.

CHARLES THOMSON was the secretary of the Continental Congress from the first session to the end. He was bitterly disappointed when he was not named to a position in the new federal government he felt was suitable and withdrew to private life, where he made a translation of the Bible. He died in 1824 at age ninety-five.

BETSY ROSS (Elizabeth Griscom Ross Ashburn Claypoole), the controversial flag maker and seamstress, remarried when her first husband John Ross, died of his injuries from an explosion. She lost her second husband, John Asburn, in 1777 when he was captured by the British while privateering at sea. He would die in one of their inhospitable prisons. Her third husband, John Claypoole, a prisoner in the same British prison, witnessed Ashburn's death. An early suitor of Betsy, he returned home bringing the news of her husband's death and in time they were wed.

Betsy Ross made additional history in Quaker annals when she and John Claypoole joined a dissident Quaker group that called itself the Free Quaker Meeting. This group built the meetinghouse at Fifth and Arch in Philadelphia, and it still stands as part of the Independence Park complex. The Free Quakers lasted until 1834, when only she—now Widow Claypoole—and Samuel Wetherill, the texile manufacturer, remained. Because it was unseemly of them to meet alone they simply locked the meetinghouse and ended the sect.

The Quakers were nearly sundered by the Revolution. In 1777, the Pennsylvania legislature demanded they give allegiance to the Revolutionists' cause or lose their citizenship. Many violated the annual meeting strictures by paying war taxes; some supported the war by raising funds and serving on committees of defense. Others joined the Continental army and fought the British. In all, 1,276 members were disowned by the Quakers. Of these, "758 for military deviations, 239 for paying taxes and fines, 125 for subscribing to loyalty tests, 69 for assisting the war effort, 43 for accepting public office and 42 for miscellaneous deviations including watching military drills and celebrating independence."

JACOB DUCHE, the first chaplain of the American Continental Congress, was branded a traitor when he re-embraced the Anglican Church. In December 1777 he sailed to England, where he became a chaplain and secretary of an asylum for female orphans in Lambeth Parish. In 1792, when the laws of banishment were rescinded, he returned to Philadelphia after suffering a paralytic stroke. There he made yet another abrupt change in his religious affiliation shortly before he died in 1798, by converting to Swedenborgianism. He was "cursed by Americans as a traitor and held in contempt by the British as a blundering fool."

WILLIAMSBURG

LORD DUNMORE worked his baleful mischief to the end. On November 7, 1775, he issued the Dunmore Proclamation, inviting the Indians to join with the slaves "for the glorious purpose of enticing them to cut their masters' throats while they are asleep." In this single stroke Dunmore had roused the entire South against him and the Crown and drove many Tories into the radicals' camp.

By February 1776, more than 500 of his battle-eager followers, living on greatly overcrowded ships, had died of fever. In the end, his trusting Loyalists of Virginia had to flee for their lives or expire, their plantations confiscated, their fortunes lost, exiled from their native land. He sent the freed "Ethiopians" who had joined his forces for promised freedom to the West Indies, where they were sold again into slavery. His destruction of Norfolk was "a crowning piece of stupidity . . . [that] . . . insured the ruin of his own friends."

Back in England, he was rewarded for his efforts by being appointed governor of the Bahamas.

LONDON

HORACE WALPOLE, the fourth earl of Orford, the inveterate parliamentary insider, and the author of some fifty volumes of letters, histories, and memoirs that give invaluable if cynical insights into eighteenth-century England, Parliament, and royal dealings, still found time to write *The Castle of Otranto,* in 1765. The first gothic novel, it is still widely read in college fiction courses. He died at eighty in 1797. His view of the world could serve as his epitaph: "The world is a comedy to those who think, a tragedy to those who feel."

JOHN MONTAGUE, the fourth earl of Sandwich, was afflicted with a mad wife and a son who was even more of a profligate than the notorious earl himself. He tried to make amends for his inveterate jobbery in the navy by working himself and his staff far into the night during the Revolution, attempting to put a viable navy to sea. He succeeded in cleaning up much of the festering corruption of the royal dockyards and was instrumental in advancing the technique of putting copper bottoms on warships, but his addiction to political patronage and his incompetence and inefficiency as an administrator ultimately contributed to the loss of the colonies. It was under Sandwich's administration that the navy had become a collection of floating wrecks, made evident when the fighting ship *Royal George* sank at anchorage after a huge piece of its bottom fell out. Many able naval officers whom George III needed to rebuild his navy refused to serve with Sandwich. Between 1774 and 1780, 60,000 men deserted from the Navy or died of disease.

Sandwich's reputation was finally ruined beyond repair when Martha Ray, his beloved mistress of some sixteen years and the mother of two of his sons, was murdered by a disappointed suitor outside the Drury Lane Theater. This event proved so shocking it soon exposed the most intimate and scandalous details of the earl's personal and governmental life. Among other accusations, he was charged with acquiring the ruined Medmenham Monastery along with some cronies and stocking it with high-priced prostitutes. He managed to hang on for a few more years despite an increasing clamor against him, but the American colonies were already lost when he was finally pushed out of office along with Lord North in 1782. He died in retirement ten years later at age seventy-four. History has labeled him the most corrupt and the most inefficient man in eighteenth-century British public life.

The Duke of Richmond, an inveterate observer of British politics, wrote Sandwich's epitaph. "I would determine not to trust Lord Sandwich for a piece of ropeyarn."

ALEXANDER WEDDERBURN, the unprincipled and opportunistic Scot who as attorney general had politically mugged Benjamin Franklin over the affair of the Hutchinson letters, managed to climb ever higher in the British establishment during the American Revolution. In 1780 he was appointed chief justice of the court of common pleas. He was raised to the peerage as Lord Loughborough and finally became lord chancellor of England. But in 1793 his career was besmirched when he was accused of maneuvering the mentally deranged George III into signing sensitive state documents. He was dismissed from office in 1801, taking with him as a consolation the title of earl of Rosslyn. Horace Walpole summed him up in three words: "a thorough knave."

Drumbeats Toward Revolution

1760

- Pitt enforces lax American customs laws to stop smuggling and raise revenue for war debt. Here begins the colonial taxation quarrel.
- George III mounts throne when George II dies.
- Massachusetts governor Bernard appoints Hutchinson chief justice, passing over James Otis Sr. Start of bitter quarrel between Hutchinson and Sam Adams.

1761

- James Otis attacks the Writs of Assistance as unconstitutional.
- Hutchinson rules that writs are legal.

1762

- Otis's "Vindication of the House of Representatives," containing all the themes, battle cries, and catch phrases of the American Revolution.

1763

- Seven Years' War ends with Peace of Paris.
- Grenville named Chancellor of the Exchequer.
- Pontiac's War; London forbids colonists from Indian lands across the Ohio.

1764

- Grenville's Sugar Act signals new imperial control over colonies. At first Americans are largely indifferent.
- Business depression causes many colonial bankruptcies.
- Currency Act prohibits colonies from issuing paper money.

1765

- Grenville's Quartering Act requires colonies to provide quarters for troops. New York, with the most troops, declares it a hidden internal tax and refuses to comply. Parliament suspends New York legislature.
- Grenville's Stamp Act.
- Patrick Henry's Virginia Resolves reject Britain's right to tax. Other colonies enact the Virginia Resolves.
- Massachusetts calls for Stamp Act Congress.
- George III fires Grenville.
- Sam Adams turns out his first mob to protest Stamp Act. Many Loyalists attacked and beaten. Mobs riot in other colonies. Stamp agents are forced to resign. Hutchinson's house destroyed by mob.
- Otis and Adams elected to Massachusetts Assembly on wave of unrest.
- Nonimportation movement grows.
- Stamp Act Congress in New York issues Declaration of Rights and Liberties.
- Daniel Dulany's pamphlet "Considerations . . ." provides arguments against future taxes.

1766

- Pitt attacks Grenville's Stamp Act, says Parliament does not have the right to tax Americans.
- Stamp Act repealed by Parliament.
- Declaratory Act affirms Britain's right to tax colonies.
- DeLancey's New York Assembly accepts Quartering Act.
- George III ejects Rockingham. "Great Commoner" Pitt takes title of earl of Chatham; popularity plunges. Gout forces him out of government.
- Townshend becomes Chancellor of the Exchequer.

1767

- Townshend budget reduces land tax in England, causing a deficit of £500,000.
- Townshend duties deliberately provoke Americans; they will raise barely a tenth of the land tax deficit—taxes on tea, glass, paper, and dyestuffs. Duties to stay in America, paying salaries of Crown officers, which takes away colonists' power over purse strings. Colonies in turmoil.
- In September Townshend, forty-two, dies of a fever.
- Massachusetts Assembly creates Circular Letter to organize intercolonial resistance to Townshend duties.
- Massachusetts Assembly passes total boycott act.
- New York forms a rump parliament to defy the Townshend Act.
- John Dickinson's "Farmer" letters demonstrate that Townshend duties are unconstitutional. Profound influence on American thinking.

1768

- Hillsborough orders Massachusetts to call back Circular Letter.
- Massachusetts Assembly refuses. It also refuses to help Royal Governor Francis Bernard collect taxes.
- Hillsborough suspends Massachusetts Assembly. Other colonies vote support of Massachusetts.

- Customs seizes Hancock's ship *Liberty*, causes riots. John Adams defends Hancock in court.
- Governor Bernard calls troops from Halifax. Sam Adams and Sons of Liberty fail to stop troops from landing as threatened, becomes laughingstock of the colonies.
- Boston refuses to quarter troops.
- Governor Bernard closes assembly.
- Sam Adams forms illegal Massachusetts assembly.
- Colonial merchants adopt nonimportation.
- Wilkes affair costs Parliament loss of colonial respect.

1769

- Hutchinson replaces Bernard as governor.
- Adams breaks into mansion of absent Bernard, finds his private papers, starts newspaper propaganda campaign.
- Adams pressures Hutchinson to move troops outside city.

1770

- Lord North takes power.
- Adams's gang enforces nonimportation agreement. Many small businessmen bankrupted.
- Hutchinson informs King George III that government in the colonies is at an end. Soldiers attacked on streets.
- Nonimportation agreement collapsing. New York merchants lead the way. Even Whigs murmur against Adams's leadership.
- Otis leads the mutiny against Sam Adams, reaffirms his loyalty to the Crown. Hancock attacks Adams for destroying Boston economy. John Adams retires from politics.
- Boston Massacre, probably engineered by Sam Adams. Hutchinson addresses crowd, averts riots. Adams demands that British troops be removed from city.
- London Massacre; troops fire on mob demonstrating for Wilkes.
- Townshend Act is repealed. Colonies celebrate.
- Parliament retains tea duty.
- Boston Massacre trials. Defense by Josiah Quincy and John Adams. In effect, juries decide there had been no Boston Massacre.

1771

- The nadir of Sam Adams's career. Nonimportation agreements collapse. John Hancock breaks with Adams, moves closer to Hutchinson who hires writers to attack Adams in Draper's *Massachusetts Gazette.* Hutchinson is winning back power for the crown.
- Otis returns to Boston politics and, denouncing Adams, takes over command again.
- Ben Franklin adds Massachusetts to the string of colonies he represents in England.
- Boston merchants, enjoying a business boom, avoid affiliation with Adams's Boston radicals. They reject violence.
- Hutchinson, to get legislation through Assembly, moves it back to Boston. A lonely victory for Adams.
- Isolated, Adams redoubles his attacks in the newspapers with dozens of articles. Flirts dangerously with treason. Seeking new issues, he is floundering.

1772

- Boston Assembly threatens secession.
- Sam Adams forms Committees of Correspondence.
- Virginia calls for thirteen permanent Committees of Correspondence. Precursor of Continental Congress.
- Sam Adams and Hancock patch up quarrel.
- Customs ship *Gaspee,* aground, attacked and burned. Britain unable to find the culprits.

1773

- Governor Hutchinson raises issue of two Parliaments, causes a firestorm on both sides of the Atlantic.
- Virginia Committee of Correspondence formed, modeled on Boston's. Beginning of intercolonial union.
- Franklin sends Hutchinson letters to Boston. Betraying Franklin with publication, Sam Adams demands Hutchinson's impeachment. Times are good and no one really cares.

- Sam Adams calls for an American congress to form an independent state, an American Commonwealth.
- Parliament passes the Tea Act. For radicals whose fortunes had been waning, tea tax is "manna from heaven."
- December 16, 1773, Boston Tea Party.

1774

- January 20, London outraged with news of the tea party.
- January 29, Wedderburn attacks Franklin in the "Cockpit" chamber over Hutchinson letters.
- Lord North passes five Coercive Acts. Boston to be closed. Annulment of Massachusetts charter. Trial in England. Quartering troops made legal. General Gage named royal governor of Massachusetts.
- Quebec Act establishes Roman Catholicism in Canada, extends Canada to the Ohio Valley. Colonists, forbidden by law to enter Ohio, are furious.
- Louis XV dies. Grandson Louis XVI becomes king.
- House of Commons refuses Massachusetts petition to remove Thomas Hutchinson as governor-general.
- May 22, Virginia House of Burgesses supports Boston with day of fasting. Governor Dunmore dissolves house.
- June 1, Gage blockades Boston.
- Governor Hutchinson sails for England and into exile.
- Sam Adams calls for a Solemn League and Covenant, the most extreme boycott. Adams nearly loses control of things when 800 merchants in Boston refuse to go along with his boycott. Other colonies suspicious of covenant. Boston merchants want to pay for tea and shut down the Committee of Correspondence.
- Three colonies call for a Continental Congress: Massachusetts, New York and Virginia.
- Gage captures 250 half barrels of powder from the Provincial Powder House; 30,000 colonists march on Boston. Gage, stunned, takes defensive measures around city.
- First Virginia Convention meets.
- Suffolk County Resolves, written under Sam Adams's direction, claim king has lost right to loyalty of subjects, sets up an independent Massachusetts Assembly, sequesters king's taxes, and calls for embargo of British goods.

- First Continental Congress meets in Philadelphia in September. Adopts Suffolk Resolves, passes nonimportation bill, petitions king with Declaration of Rights and Grievances. Adams just barely defeats Galloway's Plan of Union, calling for an American Parliament.
- Gage refuses to open Massachusetts legislature. Adams's radicals create an extralegal Provincial Congress to meet in Concord.
- Gage asks king to suspend Coercive Acts, send more troops. King rejects this.

1775

- Chatham calls for withdrawal of British troops. Parliament declines.
- Chatham offers second bill authorizing Continental Congress to tax itself. Rejected.
- Lord North's Conciliatory Plan offers colonies right to tax themselves for defense money, but still insists on Parliament's right to control colonies.
- Patrick Henry's famous ''Liberty or Death'' speech. Asserts war is inevitable. Virginia convention votes to raise a militia.
- April 19, American Revolution starts at Lexington and Concord.
- April 20, Governor Dunmore seizes Williamsburg gunpowder.
- April 23, Massachusetts Provincial Congress authorizes an army of 13,600 militia.
- May 5, Patrick Henry raises an army, forces Dunmore to pay for the confiscated gunpowder.
- May 10, Second Continental Congress opens in Philadelphia.
- May 9–10, Fort Ticonderoga captured.
- May 18, Congress shocked by news of Ticonderoga.
- May 25, Major Generals Howe, Clinton, and Burgoyne arrive in Boston.
- May 28, Dr. Warren's report on Lexington and Concord battle reaches London.
- June 5, Dunmore reads North's olive branch proposal to the burgesses. In reply, burgesses ask for the key to the powder magazine. Dunmore flees to HMS *Fowey* and calls slaves to arms against their white masters.
- June 10, General Gage's report on Lexington and Concord reaches London.

- June 15, Congress names Washington commander in chief days before Breed's Hill.
- British warships fire on Breed's Hill fortifications.
- Regulars burn Charlestown, attack Americans on Breed's Hill three times. Howe fails to pursue and take Cambridge, even though he has plenty of fresh troops and the Americans are exhausted and out of powder. British casualties—42 percent.
- Benedict Arnold's attack in Quebec fails.
- July 8, Congress sends Olive Branch Petition to King George.
- August 23, King George declares war on colonies.
- September, General Howe replaces Gage, who is recalled to England.
- King George leases 29,000 Hessian mercenaries.

Endnotes

BOSTON

p. 3 . . . along the road to Concord. Much of the material for this section was drawn from three sources. Professor Fischer's exhaustive work, *Paul Revere's Ride;* Frothingham, whose work includes many eyewitness accounts; and, for the British view, Brendan Morrisey's *Boston, 1775.*

p. 7 . . . tarred and feathered others. Fischer 73

p. 7 . . . hysterical rebelliousness. Schlesinger 92

p. 8 . . . Charlestown Congregationalist Church. Fischer 106

p. 9 . . . a durable legend it would create. Fischer 121

p. 10 . . . too "unbelievably stupid" to be devious. Fischer 106

p. 13 . . . lay a huge trunk. Fischer 181

p. 19 . . . the same terrifying symptoms. Fischer 165–6

p. 19 . . . "as long as he could stand." Morrissey 25

p. 20 . . . drunk for a copper or two." Fischer 67

p. 20 . . . than the Yankies will." Fischer 67

p. 20 . . . "half of them." Fischer 67

p. 21 . . . the Anniversary of the Battle of Lexington!" *Massachusetts Spy,* May 3, 1775

p. 21 . . . What to do? Fischer 101–3

p. 22 . . . among the dark streets of Boston's North End. Fischer 101–3

p. 22 . . . the shore of Charlestown. Langguth 231

p. 24 . . . in three hours. Fischer 141

p. 25 . . . on two very tired mounts. Fischer 143

p. 29 . . . let it begin here." Commager and Morris 70

p. 31 . . . Bunker Hill. Fischer 195–7, 403

p. 31 . . . and into the woods." Commager and Morris 71
p. 32 . . . we made off with the trunk." Commager and Morris 69
p. 32 . . . probably from a pistol. Fischer 193
p. 32 . . . there was an accident." Fischer 194
p. 32 . . . orders from Samuel Adams." Morrissey 36
p. 33 . . . only in home commands." Morrissey caption p. 20
p. 33 . . . set everything to rights." Miller, *Origins* 400
p. 34 . . . cut him to pieces." Miller, *Origins* 400
p. 34 . . . the road to Concord. Fischer 200
p. 34 . . . "For America, I mean." Fischer 183
p. 34 . . . to capture arms. Fischer 143–44
p. 35 . . . curving back toward Boston. Fischer 148
p. 35 . . . see her husband alive again. Fischer 165–66
p. 35 . . . inch by inch." Draper 492
p. 36 . . . in the early sunshine." Frothingham 66
p. 36 . . . the high ground. Gross 123
p. 37 . . . rum to the regulars. Chastellux 482. Jones Inn, which stood on the north side of the roadway, now Main Street, no longer exists.
p. 37 . . . searched every home. Frothingham 66
p. 37 . . . ill afford the loss. Gross 121–22
p. 37 . . . nothing but flour. Gross 123
p. 38 . . . the town's liberty pole. Frothingham 66
p. 38 . . . Barrett farms. Gross 123
p. 39 . . . for God's sake, fire!" Gross 124
p. 40 . . . uncivilized Savages." Gross 127
p. 40 . . . a scandal in Europe. Fischer 218, 287
p. 42 . . . a few days later. Fischer 251
p. 43 . . . retreating redcoats. Langguth 248ff
p. 47 . . . were put to death." Commager and Morris 74
p. 47 . . . which followed us." Commager and Morris 87
p. 47 . . . column had passed." Commager and Morris 87
p. 47 . . . numerous descendants. Fischer 257
p. 47 . . . in an instant." Fischer 258
p. 48 . . . lay concealed in them." Commager and Morris 87
p. 49 . . . masterpiece. Fischer 247
p. 49 . . . out of range. National Park Service, "Minute Men"
p. 49 . . . had been brilliant. Commager and Morris 89
p. 49 . . . past seven o'clock." Fischer 262
p. 50 . . . 2 percent. National Park Service, "Minute Men"
p. 50 . . . 4 missing. Gross 130
p. 50 . . . from my side." Smith 488
p. 50 . . . into Boston Gaol. Fischer 273
p. 50 . . . prisons of both sides." Lancaster 163
p. 51 . . . political engine!" Bowen 337
p. 51 . . . gathering darkness." Fischer 261
p. 52 . . . extensive country." Miller, *Origins* 357
p. 52 . . . her just authority." Miller, *Origins* 356–57
p. 53 . . . being destroyed. Miller, *Origins* 149
p. 55 . . . Whigs left Boston. Langguth 307
p. 55 . . . or defensively. Draper 477

p. 58 . . . was penniless exile. Miller, *Origins* 287
p. 58 . . . obey his rulings. Calhoon 273–74
p. 59 . . . departed for Boston. Callahan, *Royal Radiers* 56–7
p. 59 . . . a chariot." Tyler 358
p. 59 . . . in the Province's history." Bowen 512
p. 60 . . . to this day. Tyler 356
p. 60 . . . one-third of mankind." Tyler 357
p. 60 . . . to leave her behind. Callahan, *Royal Raiders* 54
p. 60 . . . cord wood stick." Callahan, *Royal Raiders* 42
p. 61 . . . torn off his back." Ann Hulton, *Letters of a Loyalist Lady,* Callahan 44
p. 61 . . . with . . . happiness." Calhoon 258
p. 61 . . . January 1776. Calhoon 257–58
p. 61 . . . slender ranks. Lancaster 111
p. 62 . . . hastily resigned. Calhoon 275
p. 62 . . . Ruggles family. Calhoon 276
p. 62 . . . a Tory. Calhoon 342
p. 62 . . . tarred and feathered. Calhoon 277
p. 62 . . . Castle William. Calhoon 277
p. 62 . . . his throat. Calhoon 282–83
p. 63 . . . never emerged. Tyler 110
p. 63 . . . for a full account of the attack on the Hutchinson estate, see Bowen 271–75.
p. 66 . . . dine on *rats.*" Commager and Morris 147–48
p. 66 . . . Robinson." Commager and Morris 148
p. 66 . . . the clapboards." Draper 504
p. 68 . . . Empire. Fischer 279
p. 68 . . . great man. Langguth 146
p. 69 . . . not to marry early." Smith, *Adams* 39
p. 69 . . . far behind his peers. Langguth 144
p. 69 . . . with child." Langguth 148
p. 71 . . . Vermin thou art!" Bowen 220
p. 71 . . . inalterable. Bowen 420
p. 72 . . . better than I can." Langguth 345
p. 72 . . . from our seats." Lancaster 149
p. 72 . . . all the Day long." Bowen xv
p. 74 . . . Dr. Warren warned him. Langguth 253
p. 75 . . . forever. Fischer 277
p. 75 . . . all of them radical. Mott 95
p. 75 . . . Mountebank." Schlesinger 132
p. 76 . . . monster in government." Schlesinger 138–89
p. 76 . . . skunk." Langguth 35
p. 76 . . . murderers." Schlesinger 138, 145
p. 76 . . . *sweepers.* Fischer 74
p. 76 . . . Libellers." Schlesinger 98
p. 77 . . . a FREE PRESS." Schlesinger 97–98
p. 79 . . . commodious barracks." Frothingham 102
p. 81 . . . same price." Langguth 289
p. 82 . . . John Thomas. Lancaster 99
p. 83 . . . Tradesmen." Miller, *Origins* 400
p. 83 . . . General Wolfe." Miller, *Origins* 400
p. 84 . . . "workmanlike, careful, unimaginative." Cook 80

p. 84 . . . £600 a year. Fischer 36, 75
p. 84 . . . in America. Fischer 39
p. 85 . . . against Boston. Fischer 38
p. 85 . . . in that city. Fischer 38
p. 87 . . . with force." Fischer 76
p. 88 . . . threw it away. Fischer 117
p. 88 . . . bullet molds. Fischer 124, 126
p. 89 . . . five crucial hours. Fischer 233
p. 89 . . . interruption at Lexington." Commager and Morris 74
p. 89 . . . half a mile." Fischer 263
p. 89 . . . possible to be." Commager and Morris 74
p. 89 . . . completely." Langguth 229
p. 89 . . . yesterday." Fischer 264
p. 89 . . . supposed them to be." Fischer 264
p. 90 . . . officers and all." Plumb 129
p. 90 . . . run away." Fischer 154
p. 91 . . . intrepidity." Commager and Morris 76–77
p. 92 . . . countrymen." Fischer 96
p. 92 . . . out of doors. Fischer 267
p. 95 . . . shocking manner." Miller, *Origins* 409
p. 96 . . . used to arms." Draper 73
p. 96 . . . this contest." Callahan, *Royal Raiders* 7
p. 97 . . . submit to." Langguth 263
p. 97 . . . Seven Years' War. Miller, *Adams* 99
p. 99 . . . plucked gawky." Miller, *Adams* 100
p. 99 . . . for Adams. Canfield 25
p. 99 . . . bills of its leader." Miller, *Adams* 100
p. 101 . . . the Sons of Liberty. Langguth 205

NEW YORK

p. 106 . . . twenty-four miles. Callahan, *Royal Raiders* 71
p. 107 . . . muskets. Champagne, 82
p. 107 . . . [pro-Loyalist] sentiments." Champagne 83
p. 107 . . . Harvard." Bowen 463
p. 109 . . . him adrift." Callahan, *Royal Raiders* 139–40
p. 109 . . . little fool!" Callahan, *Royal Raiders* 139
p. 110 . . . at any age." *Encyclopaedia Britannica* vol 11 p 121
p. 110 . . . London gentleman. Mott 85
p. 111 . . . dregs of mankind." Callahan, *Royal Raiders*, 59
p. 112 . . . this step." Callahan, *Royal Raiders*, 65
p. 112 . . . skunk." Calhoon 245
p. 113 . . . contraband" Calhoon 247
p. 113 . . . horrid tragedy." Tyler 344
p. 114 . . . embargo. Tyler 344
p. 116 . . . among ourselves." Kierner 206
p. 122 . . . every way," Champagne 39
p. 124 . . . "exploded cracker." Champagne 77

PHILADELPHIA

p. 132 . . . saw one. Bowen 467
p. 132 . . . every side. Miller, *Origins* 379
p. 132 . . . well warned. Bowen 482
p. 133 . . . Act itself." Canfield 60
p. 134 . . . in their chairs. Illick 281
p. 136 . . . silence. Galvin 285
p. 137 . . . adoption. Tyler 373
p. 140 . . . away. Bowen 503
p. 140 . . . the ocean. Bowen 603
p. 141 . . . accord. Miller, *Origins* 392
p. 141 . . . individual." Miller, *Origins* 388
p. 142 . . . demagogues. Tyler 372
p. 143 . . . vain." Bowen footnote 496
p. 143 . . . [my] own door." Callahan, *Royal Raiders* 97–98
p. 143 . . . entered Philadelphia. Tradition and old records say the rider was Israel Bissell. Professor Fischer is doubtful. He believes Bissell ended his ride in New York City.
p. 144 . . . of the Revolution." Tyler 234
p. 145 . . . and tumult." Tyler 235
p. 145 . . . Bostonian." Schlesinger 88–90
p. 145 . . . "To the Farmer." Illick 260
p. 145 . . . and infamous." Langguth 266
p. 148 . . . Continental Congress. Bowen 465
p. 149 . . . papers and magazines." Kelley 167
p. 150 . . . bowling in town. Kelley 171
p. 150 . . . and a long etc." Bowen 466
p. 150 . . . porter, beer . . . " Bowen 469
p. 151 . . . Peale. Illick 270
p. 156 . . . our independence. . . ." Rush 112
p. 158 . . . worthy, young man." Tyler 452–55
p. 158 . . . his soul." Paine 26
p. 158 . . . and fearless Whig." Rush 114
p. 162 . . . copies. Langguth 341
p. 162 . . . his own. Langguth 341
p. 162 . . . crapulous mass." Bailyn, *Faces of the Revolution* 71
p. 162 . . . striking a style. . . ." Langguth 340
p. 162 . . . English language." Bailyn, *Faces of the Revolution* 71
p. 163 . . . social equal. Kelley 177
p. 164 . . . lawn sleeves for himself." Bowen, *Adams* 465
p. 165 . . . of engagements." Norton, *Liberty's Daughters* 7
p. 165 . . . outside world. Drinker's diary 63
p. 165 . . . left Philadelphia. Norton, *Liberty's Daughters* 203
p. 166 . . . deluding young Girls." Norton, *Liberty's Daughters* 205
p. 167 . . . organization. Norton, *Liberty's Daughters* 156
p. 167 . . . with impunity." Norton, *Liberty's Daughters* 50
p. 167 . . . new code. Norton, *Liberty's Daughters* 163
p. 168 . . . secondary." Norton, *Liberty's Daughters* 61
p. 168 . . . outside the home. Norton, *Liberty's Daughters* XIV

p. 168 . . . at 40 or 45." Canfield 37
p. 168 . . . "my farming business." Norton *Liberty's Daughters* 219
p. 169 . . . equal command." Norton, *Liberty's Daughters* 223–34
p. 170 . . . of my heart." Norton, *Liberty's Daughters* 217–18
p. 170 . . . short Existence." Norton, *Liberty's Daughters* 64
p. 170 . . . thy own." Norton, *Liberty's Daughters* 219
p. 171 . . . restored." Noel Hume 45
p. 172 . . . suitable time." Frothingham 94
p. 172 . . . 38 blunderbusses." Frothingham 95

WILLIAMSBURG

p. 178 . . . never still." Blanco 911
p. 178 . . . odd genius" Blanco 912
p. 181 . . . night." Kelley 181
p. 182 . . . white man." Calhoon 221
p. 182 . . . everything wicked," he said. Tyler 460
p. 183 . . . blow your brains out." Calhoon 219
p. 183 . . . from the people." Calhoon 232
p. 188 . . . considerable estate." Noel Hume 20
p. 188 . . . serve in Virginia. Noel Hume 19–20
p. 188 . . . better husband." Noel Hume 23–24
p. 190 . . . me death!" Noel Hume 114. The Henry speech is abstracted from the account given by Henry's earliest biographer, William Wirt, in 1817. For a fuller version see Noel Hume 112ff.
p. 191 . . . militant Americans." Noel Hume 116
p. 196 . . . colonies. Miller, *Origins* 126.
p. 197 . . . to prevent." Noel Hume 178
p. 198 . . . commerce." Noel Hume 54
p. 199 . . . Lord Dunmore." Callahan, *Royal Raiders* 83
p. 199 . . . to obedience." Noel Hume 164
p. 199 . . . I can reach." Noel Hume 163
p. 202 . . . necessary luxury." *Encyclopaedia Britannica* 12:987
p. 202 . . . personal quarrel." *Encyclopaedia Britannica* 12:987
p. 203 . . . by the hour." *Encyclopaedia Britannica* 12:987
p. 203 . . . childbirth. *Encyclopaedia Britannica* 12:987
p. 203 . . . conversation." *Encyclopaedia Britannica* 12:987
p. 204 . . . tyrants." Bronowski and Mazlish 377
p. 205 . . . United States." Bailyn, *Faces* 27
p. 208 . . . "wear off." Riley, xvii

PREPARING FOR WAR

p. 215 . . . more than one." Steele 13
p. 217 . . . this country." Frothingham 107
p. 221 . . . up their escritoires." Cook 182
p. 221 . . . Master of Mischief." Cook 185

p. 221 . . . living." Callahan, *Royal Raiders* 96
p. 221 . . . open rebellion." Cook 202
p. 225 . . . not known. Frothingham 108
p. 235 . . . inquiries." Langguth 148
p. 237 . . . born." Bowen 217
p. 238 . . . Adams. Miller, *Sam Adams* 40
p. 242 . . . purposes." Canfield 27
p. 243 . . . at an end." Lewis 48

LONDON

p. 248 . . . hangings." Tuchman, *March* 135
p. 248 . . . on your souls. Brodie 103
p. 248 . . . amazement." Cook 15
p. 249 . . . content." Lancaster 41
p. 250 . . . smaller. Cook 138–39
p. 250 . . . smugglers." Lancaster 13
p. 251 . . . family inside. Clarke 52
p. 252 . . . some minutes." Smith, *John Adams* 494
p. 253 . . . monster again." Brooke 15
p. 255 . . . dominion." Galvin 280
p. 255 . . . to him." Hibbert 150
p. 256 . . . be a king!" Tuchman, *March,* 138
p. 258 . . . barbarity." Clarke 76
p. 259 . . . stank." Clarke 163
p. 260 . . . counting house clerk." Cook 52
p. 260 . . . moorings. Cook 56; Tuchman, *March,* 133
p. 260 . . . Mr. Pitt's Victories." Cook 56
p. 260 . . . his failing." Brooke 107
p. 260 . . . hour more." Brooke 108
p. 261 . . . dearly." Cook 52
p. 261 . . . indifference." Tuchman, *March* 142
p. 262 . . . protection." Miller, *Origins* 287
p. 262 . . . enslaved." Cooke 86
p. 263 . . . complexion." Tuchman, *March* 143
p. 263 . . . politicians in England." Miller, *Origins* 241
p. 263 . . . attention." Tuchman, *March* 169–70
p. 263 . . . fellow in England." Tuchman, *March* 170
p. 264 . . . faculty." Tuchman, *March* 170
p. 266 . . . heavy drinking." Cook 15
p. 267 . . . impenetrable secret." Tuchman, *March* 132
p. 267 . . . estates." Tuchman, *March* 137
p. 268 . . . disgrace." Tuchman, *March* 140
p. 268 . . . [five] seats." Tuchman, *March* 141
p. 269 . . . and corrupt." Tuchman, *March* 143
p. 269 . . . high stakes." Lancaster 15
p. 270 . . . leading them." Lancaster 15
p. 271 . . . the Ministry." Smith, *John Adams* 494
p. 271 . . . Boston!" Smith, *John Adams* 494

p. 271 . . . sadly wrote. Tuchman, *March* 203
p. 271 . . . commenced!" Commager and Morris 70
p. 272 . . . Concord." Smith, *A New Age* 494
p. 272 . . . New England." Smith, *A New Age* 495
p. 272 . . . claim a victory." Fischer 275
p. 272 . . . two shores. Tuchman, *March* 220
p. 272 . . . keep up with them." Miller, *Origins* 410
p. 272 . . . all of America. Draper 418
p. 273 . . . against them." Canfield 59
p. 273 . . . cats." Smith, *John Adams* 451
p. 273 . . . agreements." Smith, *John Adams* 451
p. 273 . . . ours." Tuchman, *March* 164
p. 273 . . . cannot exert." Smith, *John Adams* 192
p. 273 . . . your policy." Tuchman, *March* 200
p. 274 . . . any Englishman." Tuchman, *March* 198
p. 274 . . . abyss!" Tuchman, *March* 212
p. 274 . . . mind." Calhoon 41
p. 274 . . . desperate war." Tuchman, *March* 213
p. 274 . . . common sense." Cook 234
p. 274 . . . a victory. Tuchman, *March* 207
p. 275 . . . by management." Commager and Morris 14
p. 275 . . . to be found." Cook 189
p. 275 . . . the sword." Tuchman, *March* 199
p. 275 . . . conquer America." Tuchman, *March* 199
p. 276 . . . tax America." Cook 190
p. 276 . . . behind you . . . ?" Tuchman, *March* 204
p. 276 . . . 68 to 18. Tuchman, *March* 204
p. 277 . . . undiminished." Tuchman, *March* 204
p. 277 . . . and corruption." Tuchman, *March* 205
p. 277 . . . the 19th instant." Commager and Morris 14
p. 278 . . . intrepidity." Fischer 276
p. 278 . . . and Clinton. Fischer 276
p. 279 . . . submit to them." Cook 190, Commager and Morris 13–14
p. 279 . . . it deserves." Cook 221
p. 280 . . . New Empire." Lancaster 23
p. 280 . . . enemy." Smith, *John Adams* 494
p. 280 . . . submit." Smith, *John Adams* 495

AFTERWARD

p. 283 . . . well enough," Boatner, *Dictionary* 405
p. 285 . . . "stupid beyond belief." Fischer 86
p. 286 . . . newspapers. Mott 78–79, Schlesinger 258
p. 287 . . . it was still there. Fischer 121
p. 291 . . . blundering fool." Boatner, *Encyclopedia* 338
p. 292 . . . own friends." Callahan, *Royal Raiders* 85
p. 292 . . . those who feel." Baugh 1077
p. 292 . . . disease. Lancaster 276
p. 293 . . . ropeyarn." Lancaster 276

Bibliography

Adams, John. *Diary and Autobiography of John Adams.* Edited by L. H. Butterfield. 4 vols. (Harvard University Press, 1961).

Andrews, Charles. *The Colonial Background of the American Revolution* (Yale University Press, 1924).

Bailyn, Bernard. *Faces of the Revolution: Personalities and Themes in the Struggle for American Independence* (Knopf, 1990).

———. *The Ordeal of Thomas Hutchinson* (Harvard University Press, 1974).

Bakeless, John. *Turncoats, Traitors and Heroes* (Lippincott, 1959).

Baugh, Albert C., ed. *A Literary History of England* (Appleton-Century-Crofts, 1948).

Blanco, Richard L., ed. *The American Revolution 1775–1783: An Encyclopedia,* Vol. 1 (Garland Publishers, 1993).

Boatner, Mark Mayo. *Cassell's Biographical Dictionary of the American War of Independence, 1763–1783* (Cassell, 1973).

———. *Encyclopedia of the American Revolution* (D. McKay, 1966).

———. *Landmarks of the American Revolution* (Hawthorn Books, 1975).

Boswell, James. *Boswell's London Journal, 1762–63* (McGraw-Hill, 1950).

———. *Life of Johnson* (Oxford University Press, 1966).

———. *Journals of James Boswell, 1762–1795* (Yale University Press, 1970).

———. *Boswell in Extremis 1776–1778* (McGraw-Hill, 1970).

Bowen, Catherine Drinker. *John Adams and the American Revolution* (Little, Brown, 1950).

Brodie, Fawn. *Thomas Jefferson: An Intimate Biography* (W. W. Norton, 1974).

Bronowski, J., and Mazlish, Bruce. *The Western Intellectual Tradition: From Leonardo to Hegel* (Dorset Press, 1986).

Brooke, John. *King George III* (McGraw-Hill, 1972).

Brown, Alice. *Mercy Warren* (Charles Scribner's Sons, 1896).

Calhoon, Robert McCluer. *The Loyalists in the Revolutionary War 1760–1781* (Harcourt Brace Janovich, 1973).

Callahan, North. *Royal Raiders: The Tories in the American Revolution* (Bobbs Merrill, 1963).

———. *Flight from the Republic* (Bobbs Merrill, 1967).

Canfield, Cass. *Sam Adams's Revolution, 1765–1776* (Harper and Row, 1976).

Casey, William J. *Where and How the War Was Fought* (William Morrow, 1976).

Champagne, Roger. *Alexander McDougal and the American Revolution in New York* (Union College Press, 1975).

Chastellux, Marquis de. *Travels in North America in the Years 1780, 1781 and 1782* (University of North Carolina Press, 1963).

Chidsey, Donald Barr. *The Loyalists* (Crown Publishers, 1973).

Clarke, John. *The Life and Times of George III* (Book Club Associates, London, 1972).

Commager, Henry Steele, and Morris, Richard B., eds. *The Spirit of 'Seventy-Six* (Harper and Row, 1958).

Cook, Don. *The Long Fuse: How England Lost the American Colonies 1760–1775* (The Atlantic Monthly Press, 1995).

Cully, Margo, ed. *A Day at a Time: The Diary Literature of American Women from 1764 to the Present* (Feminist Press, 1985).

Curson, Julie P. *A Guide's Guide to Philadelphia* (Curson House, 1992).

Davidson, Philip. *Propaganda and the American Revolution 1763–1783* (University of North Carolina Press, 1941).

Davis, Burke. *Heroes of the American Revolution* (Random House, 1971).

des Cognets, Louis, Jr. *Black Sheep and Heroes of the American Revolution* (Princeton University Press, 1965).

Dictionary of National Biography (Oxford University Press, 1963–65).

Donovan, Frank. *The Women in Their Lives: The Distaff Side of the Founding Fathers* (Dodd, Mead and Company, 1966).

Draper, Theodore. *A Struggle for Power: The American Revolution* (Times Books/Random House, 1996).

Drinker, Elizabeth Sandwith. *Extracts from the Journal of Elizabeth Sandwith Drinker, 1759–1807* (J. B. Lippincott, 1889).

Dupuy, Trevor, and Hammerman, Gay M., eds. *People and Events of the American Revolution* (Bowker, 1974).

Encyclopaedia Britannica, 1956.

Ferling, John. *John Adams: A Life* (University of Tennessee Press, 1992).

———. *The Loyalist Mind: Joseph Galloway and the American Revolution* (Penn State Press, 1977).

Ferris, Robert G., and Morris, Richard E. *The Signers of the Declaration of Independence* (National Park Service Interpretive Publications, 1982).

Fischer, David Hackett. *Paul Revere's Ride* (Oxford University Press, 1994).

Franklin, Benjamin. *Autobiography* (Garden City Publishing Co., 1916).

French, Allen. *The Siege of Boston* (Macmillan, 1911).

Frothingham, Richard Jr. *History of the Siege of Boston, and of the Battles of Lexington, Concord and Bunker Hill* (Charles C. Little and James Brown, 1849).

Galvin, John R. *Three Men of Boston* (Thomas Y. Crowell Company, 1976).

Grafton, John. *The American Revolution: A Picture Source Book* (Dover Publications, 1975).

Greene, Jack P., and Pole, J. R. *The Blackwell Encyclopedia of the American Revolution* (Basil Blackwell Inc., 1991).

Gross, Robert A. *The Minutemen and Their World* (Hill and Wang, 1976).

Hamm, Margherita. *Famous Families of New York* (Putnam, 1902).

Hart, Albert Bushnell. *American History Told by Contemporaries, 1897–1929* (Macmillan Company, 1898).

Hezekiah, Niles. *Chronicles of the American Revolution* (Grosset and Dunlap, 1965).

Hibbert, Christopher. *The Story of England* (Phaidon Press Ltd., 1992).

Illick, Joseph E. *Colonial Pennsylvania: A History* (Charles Scribner's Sons, 1976).

Kashatus, William C. III. *Historic Philadelphia: The City, Symbols and Patriots 1681–1800* (University Press of America, 1992).

Kelley, Joseph J. Jr. *Life and Times in Colonial Philadelphia* (Stackpole Books, 1973).

Kierner, Cynthia A. *Traders and Gentlefolk: The Livingstons of New York, 1675–1790* (Cornell University Press, 1992).

Kipping, Ernst. *The Hessian View of the American Revolution, 1776–1783* (Philip Freneau Press, 1971).

Lancaster, Bruce. *The American Heritage History of the American Revolution* (American Heritage, 1971).

Langford, Paul. *A Polite and Commercial People: England 1727–1783.* The New Oxford History of England (Clarendon Press, 1989).

Langguth, A. J. *Patriots: The Men Who Started the American Revolution* (Simon and Schuster, 1988).

Leonard, Eugenie A; Drinker, Sophie; and Holden, Miriam Y. *The American Woman in Colonial and Revolutionary Times, 1565–1800* (University of Pennsylvania Press, 1962).

Lewis, Paul. *The Grand Incendiary: A Biography of Sam Adams* (Dial Press, 1973).

Lopez, Claude-Anne. *Benjamin Franklin's "Good House": The Story of Franklin Court* (Division of Publications, National Park Service, U.S. Department of the Interior, 1981).

Middlekauff, Robert. *The Glorious Cause: The American Revolution.* Oxford History of the United States (Oxford University Press, 1982).

Miller, John C. *Origins of the American Revolution* (Little, Brown, 1943).

———. *Sam Adams, Pioneer in Propaganda* (Little, Brown, 1936).

Montross, Lynn. *The Reluctant Rebels: The Story of the Continental Congress* (Harper and Brothers, 1950).

Morris, Richard B. *The American Revolution Reconsidered* (Harper and Row, 1967).

———. *Independence: A Guide to Independence National Historical Park* (Division of Publications, National Park Service, U.S. Department of the Interior, 1982).

Morrissey, Brendan. *Boston, 1775: The Shot Heard Around the World.* Osprey Military Campaign Series (1993).

Mott, Frank Luther. *American Journalism: A History of Newspapers in the United States Through 250 Years, 1690–1940* (Macmillan Company, 1949).

Namier, Sir Lewis Bernstein. *England in the Age of the American Revolution* (St. Martin's Press, 1961).

National Park Service, Division of Publications, U.S. Department of the Interior, "Minute Men," 1993.

Nettles, Curtis P. *The Roots of American Civilization* (Appleton-Century-Crofts, 1938).

Newcomb, Benjamin H. *Franklin and Galloway: A Political Partnership* (Yale University Press, 1972).

Noel Hume, Ivor. *1775: Another Part of the Field* (Knopf, 1966).

Norton, Mary Beth. *The British Americans: Loyalist Exiles in England, 1774–1789* (Little, Brown, 1972).

———. *Liberty's Daughters: The Colonial Experience of American Women, 1750–1800* (Little, Brown, 1980).

Paine, Thomas. *Common Sense* (Dover Publications, 1997).

Pearson, Michael. *Those Damned Rebels: The American Revolution as Seen Through British Eyes* (G. P. Putnam's Sons, 1972).

Plumb, J. H. *England in the Eighteenth Century,* The Pelican History of England v. 7. (Penguin Books, 1950.)

Purcell, L. Edward. *Who Was Who in the American Revolution* (Facts on File, 1993).

Riley, Edward Miles, ed. *The Journal of John Harrower: An Indentured Servant in the Colony of Virginia, 1773–1776* (Colonial Williamsburg, 1963).

Rush, Benjamin. *Autobiography* (Princeton University Press, 1948).

Ryerson, Richard A. *The Revolution Is Now Begun: The Radical Committees of Philadelphia, 1765–1776* (University of Pennsylvania Press, 1978).

Sabine, L. *Loyalists of the American Revolution* (Genealogical Publishing Co., 1979).

Schappes, Morris U. *The Jews in the United States* (Citadel Press, 1958).

Scheer, George, and Rankin, Hugh F. *Rebels and Redcoats* (Da Capo Press, 1987).

Scheide, John H. "The Lexington Alarm," *American Antiquarian Society Proceedings* 50 (1940):63.

Schlesinger, Arthur M. *Prelude to Independence: The Newspaper War on Britain, 1764–1776* (Knopf, 1958).

Shy, John. *A People Numerous and Armed* (University of Michigan Press, 1990).

———. *Toward Lexington: The Role of the British Army in the Coming of the American Revolution* (Princeton University Press, 1965).

Smith, Page. *A New Age Now Begins.* 2 vols. (McGraw-Hill, 1976).

———. *John Adams* (Doubleday, 1962).

Steele, Richard C. *Isaiah Thomas* (Worcester Historical Museum, 1981).

Tebbel, John. *Turning the World Upside Down: Inside the American Revolution* (Orion Books, 1993).

Tuchman, Barbara, *The First Salute* (Knopf, 1988).

———. *The Road to Folly* (Knopf, 1984).

Tucker, Glenn. *Mad Anthony Wayne and the New Nation* (Stackpole Books, 1973).

Tyler, Moses Coit. *The Literary History of the American Revolution.* 2 vols. (Frederick Ungar, 1957).

Winston, Alexander. "Firebrand of the Revolution," *American Heritage,* April 1967.

Wolf, Edwin II, and Whiteman, Maxwell. *The History of the Jews in Philadelphia from Colonial Times to the Age of Jackson* (Jewish Publication Society of America, 1957).

Wright, Esmond, ed. *A Tug of Loyalties: Anglo American Relations, 1765–85* (University of London/Athlone Press, 1975).

Young, Philip. *Revolutionary Ladies* (Knopf, 1977).

Index